Sustainable Corrosion Inhibitors

Edited by

Inamuddin[1], Mohd Imran Ahamed[2], Mohammad Luqman[3] and Tariq Altalhi[4]

[1]Department of Applied Chemistry, Zakir Husain College of Engineering and Technology, Faculty of Engineering and Technology, Aligarh Muslim University, Aligarh-202002, India

[2]Department of Chemistry, Faculty of Science, Aligarh Muslim University, Aligarh-202002, India

[3]Chemical Engineering Department, College of Engineering, Taibah University, Yanbu Al-Bahr-83, Al-Bandar District, Postal Code: 41911, Kingdom of Saudi Arabia

[4]Department of Chemistry, Collage of Science, Taif University, P.O. Box 11099, Taif 21944, Saudi Arabia

Published by **Materials Research Forum LLC**
Millersville, PA 17551, USA

Published as part of the book series
Materials Research Foundations
Volume 107 (2021)
ISSN 2471-8890 (Print)
ISSN 2471-8904 (Online)

Print ISBN 978-1-64490-148-9
eBook ISBN 978-1-64490-149-6

Distributed worldwide by

Materials Research Forum LLC
105 Springdale Lane
Millersville, PA 17551
USA
https://www.mrforum.com

Manufactured in the United States of America
10 9 8 7 6 5 4 3 2 1

Table of Contents

Preface

Corrosion of structures and equipment made from metals and alloys pose a very serious problem to the majority of industries. It costs billions of dollars to the economy worldwide. Also, it endangers lives of the workers, and affects the environment badly in case of accidents. To face and solve this challenge, material scientists have worked very hard, and developed many corrosion inhibitors and related techniques. The majority of synthesized corrosion inhibitors are toxic to the environment and/or expensive. Owing to a great focus on security, sustainability and environment, researchers are trying their level best these days to find cost-effective, environmentally friendly, renewable, and efficient corrosion inhibitor.

This book presents an overview of theories and prevention techniques behind the corrosion process. It also deals with various types of green corrosion inhibitors and their applications in numerous industries and environments. It covers the current-status, opportunities, and challenges of green corrosion inhibitors and technologies. The book will serve as an excellent reference for postgraduate students, researchers, surface scientists, material chemists, scientists, and engineers.

Key features:

- Presents an overview of theories and prevention techniques behind corrosion process
- Contributions from leading researchers in the field of corrosion science and technology
- Presents industrial applications of green corrosion inhibitors

Summaries

Chapter 1 deals with the limitations of conventional corrosion inhibitors, thereby, opening the door for different types of green corrosion inhibitors. Different sources of these inhibitors and their applicability are also presented via different works performed worldwide followed by their future scopes and opportunities. The challenges of using these inhibitors have also been highlighted.

Chapter 2 focuses on various materials that were used as corrosion inhibitors for the prevention of metal-based corrosion. This chapter also focuses on various properties of corrosion inhibitors such as improved capacity, clean, renewable, sustainable sources of energy and their cost-effective nature. Major attention is given to those materials which can serve as substitution to toxic materials and having high absorbing ability toward adherence on the metal surface leading to a prevention of corrosion.,

Chapter 3 discusses the application of specific organic substances derived from natural resources and under-utilized waste biomasses able to protect the metals and alloys in aggressive corrosive medium. Their eco-friendly nature and cost-effective prevention of metallic surfaces are particularly emphasised, according to the "green chemistry" concept in corrosion science.

Chapter 4 mainly deals with the inhibition nature of polysaccharides towards metals and alloys in various corrosive media including HCl, H_2SO_4, and NaCl. Various studies such as weight loss, the electrochemical, and computational analysis were performed in order to assess the effectiveness of polysaccharides as green corrosion inhibitors.

Chapter 5 discusses recent research contributions related to green organic corrosion inhibitors, particularly based on natural product materials and organic compounds.

Chapter 6 discusses the recent developments in the area of sustainable corrosion inhibitors for concretes used commercially in various industries. It also highlights the classification of corrosion inhibitors, concrete corrosion mechanisms and the requirement for green/sustainable corrosion inhibitors for the future with enhanced improvements.

Chapter 7 discusses green inhibitors to prevent, control or impede the growth of corrosion. Various advantages of green inhibitors such as eco–friendly nature, cost–effectiveness, and renewable, which are favourable over toxic synthetic corrosion inhibitors are discussed in this chapter in detail. It is shown that extracts of natural products containing alkaloids, carboxylic acids, nicotine, polyphenols, quinine, terpenes, and other functional groups possessing elements like C, N, O, S, etc., promote adsorption via forming a thin layer (coating) on metallic surface to shield the surface, and hence, prevent corrosion.

Chapter 8 discusses the application of numerous compounds as sustainable inhibitors to control the corrosion of copper and its alloys in various corrosive media. Amino acids, biopolymers and ionic liquids compounds, as well as plant extracts, are widely used as potent green inhibitors against the corrosion of different copper-based materials.

Chapter 9 is focused on the fundamentals of corrosion, corrosion inhibition, materials used for it and case studies of green inhibitors used for corrosion control in various conventional and monolithic metals. This chapter concludes that the extracts of plants are good corrosion inhibitors that could be positively used at industrial level.

Editors

Inamuddin, Mohd Imran Ahamed, Mohammad Luqman and Tariq Altalhi

Sustainable Corrosion Inhibitors
Materials Research Foundations107 (2021) 1-29

Materials Research Forum LLC
https://doi.org/10.21741/9781644901496-1

Chapter 1

Current Status, Opportunities and Challenges of Green Corrosion Inhibitors

Subhajit Kundu[1], Mahuya Das[2], Debarati Mitra[1*]

[1]Department of Chemical Technology, University of Calcutta, 92 A P C Road, Kolkata -700009, West Bengal, India

[2]Greater Kolkata College of Engineering and Management, Baruipur, South 24 Pgs, Kolkata-743367, India

* debarati.che@gmail.com

Abstract

The term 'Corrosion' is associated with the deterioration of materials. To get rid of this problem corrosion inhibitors are commonly used. Though many conventional corrosion inhibitors can efficiently reduce corrosion, their disposal often adversely affects the environment. However, there are many natural products that have the potential of reducing corrosion and they are referred to as green corrosion inhibitors. They are much safer to use and most importantly cost-effective. This chapter deals with the current development of the use of green corrosion inhibitors, their opportunities and challenges.

Keywords

Corrosion, Inhibition, Natural Products, Environment Friendly

Contents

Materials Research Forum LLC
https://doi.org/10.21741/9781644901496-1

1.   Introduction

The term "corrosion" originated from the Latin word *corrodere* meaning "gnawing to pieces". It is a very common detrimental issue of concern in our daily life [1]. In accordance with ASTM standard corrosion can be defined as the process of putrefaction of materials generally metals by chemical or electrochemical reactions between the materials and their adjacent environment [2]. From the time of industrial revolution i.e. late eighteenth century corrosion became a severe problem as most of the metals in application are exposed to environmental moisture, which leads to the moist environmental corrosion [3]. The modern corrosion science was pillared by Wanger and Traud in 1938 where they concluded that the mechanism of corrosion mostly occurred in the boundary region of metals and electrodes [4].

2. Different forms of corrosion

2.1 General corrosion

General corrosion is a very habitual term which occurs when a metal is exposed to an environment where chemically or electrochemically the outer layer of the metal becomes thin and hence is corroded. This type of corrosion can be prevented by choosing of proper materials of construction [5].

2.2 Metallurgical influenced corrosion

The fatigue strength of corrosion depends on the internal structure of the construction of materials. It includes intergranular corrosion (comparatively little grain corrosion occurred at the boundaries of the grain which results in defragmentation of the alloy and the strength is lost), dealloying or selective leaching (process of removing an element from an alloy by corrosion), sensitization, exfolitation etc. [5].

2.3 Mechanically assisted corrosion

This type of corrosion includes wear corrosion, erosion corrosion and corrosion fatigue. These occur when fluctuating or cyclic stress is applied on the materials. Corrosion fatigue is greatly influenced by composition of the solution, pH, oxygen content, temperature etc. [5].

2.4 Localized corrosion

This type of corrosion can be categorized as filiform corrosion, crevice corrosion, pitting corrosion. In pitting corrosion localized penetration occurs. This autocatalytic type of corrosion requires extended initiation period. Crevice corrosion is generally found in lap joints, holes, rivets and bolts etc. This type of corrosion occurs via hydroxylation mechanism. Filiform corrosion can be found in flanges, gaskets, underlying metals coated with paints etc. [5].

2.5 Microbiological corrosion

This type of corrosion is promoted by different types of bacteria including heterotrophic and autotrophic bacteria. This corrosion leads to dealloying, pitting as well as hydrogen embrittlement [5].

2.6 Environmentally induced corrosion

This class of corrosion includes hydrogen-induced cracking, hydrogen damage, stress-corrosion cracking, embrittlement, hot-cracking, liquid metal embrittlement, formation of hydride, solid metal induced embrittlement [5].

3. Corrosion inhibition

In the context of protection of metals with their alloys lots of techniques such as using chemicals which forms a protective layer of films, passing impressed current to protect the cathodes etc. are applied. Sometimes by the virtue of auto oxidation process corrosion can also be minimized.

To prevent the corrosion phenomenon corrosion inhibitors are very much readily used. Corrosion inhibitors are the chemicals which keep down the corrosion [6]. Generally, four types of environment are considered for the use of corrosion inhibitors:

- Water in the pH range 5-9
- Aqueous acid solutions
- Oil refining environment considering primary and secondary production
- Miscellaneous environments

The classification of the inhibitors can be done according to:

• Electrode process

Three types of processes are employed viz. process based on anodes, process based on cathodes and process based on both anodes and cathodes.

Now what happens during corrosion? Actually, metals have lot of free electrons which in solution state move towards anode which acts as an active site and then they are passed to the cathodic side which acts as an electron acceptor. In this context, any oxidizing agent promotes the acceptance of electrons. The extent of corrosion can be reduced by minimizing the cathodic and anodic reactions. Inhibitors are adsorbed on the metallic surface creating a strong barrier that slower down the oxidation reactions of corrosion [7].

• The chemical nature of the environment

Vapor-phase inhibitors, alkaline inhibitors, acidic inhibitors, neutral inhibitors.
Some works on conventional corrosion is summarized in Table 1.

Table 1.   Some earlier investigations on corrosion inhibition.

Observations	References
By the virtue of screening and electrochemical tests the test rate of decay by using inhibitor was observed.	8
Effect of stress-corrosion cracking on high-strength Al-alloys following the variation of mechanical properties was monitored.	9
Investigation on inhibition of aluminium-based alloys in caustic alkalis, chlorides and some mineral acidic media was performed.	10
Investigation on inhibition of copper and copper-based alloys in different media was observed.	11

4.   PARCOM guidelines for corrosion inhibition

Considering environmental concern, green corrosion inhibitors can be used instead of the conventional inhibitors, yet the effects of the green inhibitors on the environment must be taken into consideration. PARCOM (Paris Commission) delegated by EEC (The European Economy Community) provide some guidelines which can be summarized as follows.

PARCOM guidelines are based on three parameters: (i) toxicity, (ii) biodegradation and (iii) bioaccumulation.

4.1   Toxicity

Toxicity is measured as LC_{50} (Lethal Concentration that affects the population by 50%) as well as EC_{50} (Effective Concentration of chemicals required for which the population is affected by 50%). Generally, EC_{50} is lower than LC_{50}. Table 2 lists the standard test methods of toxicity tests with preferred species of different tropic levels [12].

Table 2.   Toxicity tests on different tropic levels.

Group	Preferred species	Test
Primary producer	*Skeletonema costatum*	72-h EC_{50}
Consumer	*Acartia tonsa*	48-h LC_{50}
Decomposer	*Corophium volutator*	10-day LC_{50}

4.2   Biodegradation

Biodegradation is an evaluation of the time during which a chemical persists in the vicinity of the system. This test is followed by the Organization for Economic Co-operation and Development (OECD) 301D test observing 28 days of biodegradation study.

4.3   Bioaccumulation

Bioaccumulation is measured by calculating the partition co-efficient of a chemical dispersed between two phases of 1-octanol and water. The chemical compounds present in inhibitors are examined by HPLC OECD 117 method.

A model is emerged by PARCOM which is termed as CHARM (Chemical Hazard Assessment and Risk Management) which contains four modules: Pre-screening, Hazard assessment, Risk analysis and Risk management.

Pre-screening uses a term "bad actors" which indicates the biodegradation level is less than 20% or the logarithmic value of the partition co-efficient is greater than 0.5 and the molecular weight is also greater than 700.

Hazard assessment is done by calculating the ratio of PEC (predicted environmental concentration) to NEC (no effect concentration). The eco-system will not be affected if NEC is equal or greater than PEC.

Risk analysis measures the probability of occurrence of harmful effects of the chemicals on the environment.

Risk management identifies the steps to be taken to reduce the harmful effects on the environment involving cost-effectiveness, unconventional chemicals and treatment process. It introduces two terms viz. best available technology (BAT) and best environmental practice (BEP).

The mammal toxicity LD_{50} values of conventional corrosion inhibitors tabulated in Table 3, point out the need for green corrosion inhibitors.

Table 3.   Mammal Toxicity LD_{50} values.

Compound	LD_{50} (mg/kg)
Propargyl alcohol	55
Hexynol	34
Cinnamaldehyde	2200
Formaldehyde	800
Dodecylpyridinium bromide	320
Naphthylmethyl quinolinium chloride	644
Nonylphenolethylene oxide surfactants	1310

The mechanism of corrosion inhibition by organic or inorganic compounds generally involves isolation of the metallic surface from its circumambient media in order to prevent/reduce the process of oxidation-reduction. Organic inhibitors form a defensive

Materials Research Forum LLC
https://doi.org/10.21741/9781644901496-1

layer to prevent corrosion on the surface of metal or alloy by adsorption. On the other hand, in case of inorganic inhibitors which generally act as anodic inhibitors a film is formed in which the atoms of the metal are bound to increase their resistance towards corrosion. The majority of the conventional corrosion inhibitors are toxic which create environmental issues after disposal. Moreover, the regulations regarding environment, limits their application also. For example, oilfield chemicals used as inhibitors that are discharged from offshore production platforms into the environment, adversely affect the marine life.

Regular corrosion inhibitors in general, are harmful for the environment. This triggered the need of the corrosion inhibitors which are eco-friendly. These anti-corrosive materials are most of the time derived from natural resources. So, these are far less hazardous to the environment upon rejection.

These inhibitors are also termed as green corrosion inhibitors which by a little dose can effectively reduce the problem of corrosion on the metals or alloys surface. During the process of adsorption of effective species i.e., naturally synthesized chemical compounds from plant extracts the rate of corrosion on metallic surface can be affected by:

• Altering the rate of reactions at the anode and/or cathode,

• Affecting the rate of diffusion of aggressive ions that interact with the metals,

• Enhancing the electrical resistance of the metallic surface by formation of a protective film.

5.   Theory of green corrosion inhibitors

The corrosion process is directly related to Gibbs' free energy. Higher the Gibbs' free energy higher will be the corrosion rate. To lower down the free energy green corrosion inhibitors can be used. Inhibitors accumulate on the surface of the active sites of the corroded metals and form a barrier against corrosion. The relative rate of corrosion is also regulated by the Pilling-Bedworth ratio, by which the status of the film surface can be identified [13]. When the ratio is less than 1, i.e., the volume of the product generated by corrosion is less than that of the volume of the metal, insufficient oxide to protect the metal is produced which means the metal is rendered non-protective. When the ratio is greater than 1, i.e. the volume of the product generated by corrosion is greater than the volume of metal, the metal is protective; however too large value of the ratio may create cracking in the oxides formed. When the ratio becomes 1, there is a good spatial match between the oxide and the metal, and the oxide is rendered protective [14].

Inhibition that occurs based on the adsorption technique is affected by many factors like type of the electrolytes used, temperature, chemical compositions of the inhibitors etc. For chemical adsorption process there is a chance of the formation of coordinate covalent bond. The green corrosion inhibition is generally done at room temperature and the inhibition efficiency is generally inversely proportional with temperature [15].

The action of the green corrosion inhibitors is also influenced by the structure of the ingredients which can be summarized as follows [16]:

- The inhibitors are generally in onium ions form which are adsorbed on the metal surface having active cathodic sites

- The plant extracts contain alkaloid bebeerines. It has O- and N- atoms carrying free electrons which facilitate to form bonds with the electrons that are freely available on the metal surfaces

- The potential cathodic process of steel can be affected by S-containing unsaturated compounds such as allyl propyl disulfide exited from many plant extracts

- In pyrrolidine, the strength of the bond between pyrrolidine and metals depends on the availability of the negative charge on the N- atoms which in turn denotes the basicity of the compounds

- The -OCH$_3$ group in alkaloid ricimine favors the interaction with the active metal surfaces

6.   Some Works on The Corrosion Prevention Using Green Corrosion Inhibitors

Nowadays, cleaner production, biodegradability, sustainable development, pollution prevention are to be implemented in industrial unit operations, synthesis and applications.

Currently several investigations have been carried out using different plant sources as inhibitors to prevent corrosion. Onion, ficus, natural honey, khella, olive, shirsh zallouh, vanillin, many oils extracted from different parts of various plants etc. are some of the examples which have been used for this purpose. Herbal extracts have many active groups which are the key factors inhibiting corrosion. Natural products such as tannin and lignin, cinchona alkaloids, eucalyptus leaves, *Lawsonia inermis Swertiaangustifolia*, *Calotropis giganta* latex, *officinalis*, *Terminalia belerica*, *Sapindustrifolianus*, Mahasudarshan churna, *Eugenia jambolans*, *Acacia conciana* are very fruitful for preventing different forms of corrosion mentioned earlier [17].

Farooqi and Quraishi used aqueous extract of *Azadirachta indica* and *Cordia latifolia* to prevent scaling in cooling towers. Moreover, it was noticed that the hardness of water did not affect the inhibitor. The inhibition process was also in the permissible limits of biochemical oxygen demand (BOD) and chemical oxygen demand (COD) [18]. In another work, the coordination chemistry behind the inhibition process was shown by Agarwaala by using porphyrins as corrosion inhibitors [19]. *Aloe vera* leaves, orange and mango peels have been used by Saleh et al. to prevent corrosion on steel in acidic media whereas Srivatsava et al. used caffeine and nicotine for corrosion prevention on steel [20]. Abdel-Gaber *et al*. used the extracts of black cumin and kidney bean green corrosion inhibitors to arrest corrosion on steel in acidic media [21]. *Hibiscus sabdariffa* extract has been used by El-Hosary *et al*. to analyze corrosion inhibition of aluminum and zinc in acidic medium. The corrosion inhibition of mild steel has been thoroughly monitored by Ananda et al. using ethanolic extract of *Ricinus communis* leaves in acid media [22]. Several works have been carried out by Hammouti *et al*. using different extracts of natural products such as jojoba oil, acetyl eugenol, *Mentha pulegium* etc. to study corrosion prevention of steel. The extract of *Rosmarinus officinalis* was applied by Kliskic *et al*.as corrosion inhibitor for aluminum alloy in chloride solution [23]. The effects of reducing sugars like fructose and mannose have been observed by Muller et al. on the alkaline corrosion of zinc and aluminium [24]. Some investigations on natural products as corrosion inhibitors were done by Zucchi and Omar and it was found that *Papaia, Azadirachta indica, Calotropis procera, Poinciana pulcherrima* etc. can act as efficient corrosion inhibitors in acidic environment [25]. The restrain of corrosion on industrial cooling towers has been surveyed by Minhaj et al. using aqueous extract of hibiscus flower [26]. Corrosion inhibition of copper has been monitored by El-Etre using natural honey [27]. Anticorrosive activity of *T. chebula* and *Emblica officinalis* was analyzed by Sanghvi et al. [28]. Abdallah used guar gum as anticorrosive agent [29]. Prevention of corrosion in acidic medium taking low-carbon steel as a case study has been investigated by Martinez and Stern suing Mimosa tannin [30]. The effect of inhibition using *Z. alatum* plant extract on mild steel corrosion in phosphoric acid medium by Weight Loss (WL) and Electrochemical Impedance Spectroscopy methods was studied by Chauhan and Gunasekaran. It was concluded that the prevention efficiency was about 88% at 70^0C [31]. Opuntia extract and berberine from Captis were investigated by Li et al. and found that due to the presence of heterocyclic constituents the corrosion inhibition activity became very much high [32]. Black pepper as corrosion inhibitor has been studied by Raja and Sethuraman by conventional WL studies, electrochemical TI and EIS, as well as by scanning electron microscopic studies and it was found that this reagent can act as very good corrosion inhibitor at high temperature through adsorption process [33]. The inhibition mechanism via the investigation of activation energy change and effect of temperature was investigated by Okafor et al. They concluded that the favorable adsorption model was Temkin isotherm [34]. Ethanol extract of *Colocasia esculenta* was used as corrosion inhibitor on mild steel corrosion by Eddy and the effects of phytochemical constituents have been analyzed along with the adsorption kinetics [35]. Polarization of mixed-type inhibitor like carboxymethyl-chitosan was observed by Cheng et al. and it has been observed that the protection efficiency of this inhibitor is inversely proportional to the rise in temperature (for a small range) [36].

The leaf extracts of *Chromolaena odorata* L. has been studied as green inhibitor for corrosion of aluminium in acidic medium by gasometric and thermometric techniques at different temperatures [37]. The feasibility study showed that this green inhibitor is bidodegradable as well as cost-effective. It has a variety of applications including anodizing and surface coating in industries. Guar Arabic, a water soluble, branched, neutral natural gum is very much popular in case of corrosion inhibition. As it contains high concentration of organic components it is used as adsorptive material in mild steel and aluminium corrosion in acidic environment.

Punica granatum extract as corrosion inhibitor was monitored by Deepa et al. [38] on brass and they found that the efficiency of inhibition is directly proportional to the increase of concentration of acid. The maximum efficiency achieved was 94.52% at 1000 ppm concentration of bio-inhibitor. On the basis of these results the physisorption mechanism was also studied.

Hitherto saps of *Accacia Arabica, Annona squamosa, Ricimus communis, Eugenia jambolans, Azadirachta Indica, Pongamia glabra etc.* have been scrutinized for analysis of reluctance of corrosion in the acidic environment taking mild steel as a case study. In addition, it has been found that some herbs like anis, black cumin, coriander can be used as potential, new, green corrosion inhibitors [39].

The study of protection of rebar rooted into concrete from premature deterioration under high alkaline environment was reviewed by Kundu et al. in the light of green inhibitors [40]. Considering the formation of two-faced oxides the effect of hydrophobicity of *Bambusa arundinacea* influences the mechanism of these inhibitors. The study of corrosion inhibition using *Vernonia amygdalina* extract green inhibitor has been carried out by Eyu et al. using a case study of reinforcement of carbon steel in concrete in the environment enriched with chlorides [41]. Green synthesis of inhibitors has been investigated by several other researchers [42]. Table 4 lists some natural products having anti-corrosive properties.

Table 4.   *Different natural products and their anti-corrosive properties.*

Serial No.	Metal	Medium	Inhibitor	Observations	Ref.
1	Carbon steel	NaCl (10 ppm) and $Al_2(SO_4)_3$ (35 ppm)	Propolis	Boiling water was used as solvent. Using inhibitor concentration 0.6 g/l inhibitor efficiency (IE) was achieved by 92.7% at 30^0C.	43
2	Aluminium	0.5 to 2M HCl	Citrus peel	Pectin was the major constituent. At 10^0C the IE was 91%.	44
3	Steel	1M HCl	Cladodes of *opuntia ficus indica*	Microwave extraction method was used which led to 94% IE.	45
4	Bronze	Simulated acid rain solution	*Salvia hispanica* seeds	Inhibition was carried out by soaking in methanol.	46
5	Mild steel	1M HCl	Leaves of henna	Boiling water was used as solvent. IE was 90.34% at 30^0C.	47
6	6063 Al-alloy	1M HCl	Grains of *peganum harmala*	Methanol was used as solvent followed by acidic/alkaline extraction. Inhibitor concentration was 0.025 g/l and IE was 91.78%.	48
7	Low carbon steel	1M HCl	*Schinopsis lorentzii* tree powder	TAPPI T204 OM-8 and ASTM 1110-96 standard was followed.	49
8	C38 steel	1M HCl	*Geissosper mum* leave	Acidic/alkaline extraction method was applied. IE was 90% 30^0C.	50
9	Carbon steel	1M HCl	*Retama monosperm a (L.) Boiss.* stems	Acidic/alkaline extraction method was applied. IE was 83% 30^0C.	51

10	Mild steel	1M HCl	Bark and leaves of *Neolamarckia cadamba*	Acidic/alkaline extraction method was applied. IE was 91% 30^0C.	52
11	Carbon steel	2M HCl	Dry olive leaves	Boiling water was used.	53
12	Mild steel	1M HCl	Watermelon peel, seeds and rind	Boiling in 1M HCl solution.	54
13	Tin	NaCl (2%), acetic acid (1%) and citric acid (5%) solution	Tomato peel wastes	Acidic/alkaline extraction method was applied. IE was 90% 30^0C.	55
14	Mild steel	1M HCl	Pennyroyal oil from *Mentha pulegium*	Weight Loss measurements, Electrochemical polarization and Electrochemical Impedance Spectroscopy was used.	56
15	Mild steel	0.5M H_2SO_4	*Artemisia herba alba*		57
16	Mild steel and copper	NaCl and SO_2 environment	Natural oil from *Cassia auriculata, Strychnos nuxvomica*		58
17	Concrete steel surface	10-23% NaOH	Magraba banana stems	Banana plant juice was used as inhibitor. Inhibition was monitored using WL method.	59
18	Mild steel	1M HCl	*Justicia gendarussa* extract	WL method, Atomic Force Microscopy, Electron Spectroscopy for chemical analysis was used which suggested that the inhibitor was mixed-type inhibitor. The isotherm was that of Langmuir.	60

19	Mild steel	0.1M H_2SO_4	Caffeic acid	WL method, Raman spectroscopy, potentiodynamic polarization techniques were applied which concluded that the inhibitor controls the anodic reaction.	61
20	Carbon steel	1M HCl	Aqueous extract of passion fruits and cashew peels	WL and EIS were used. The adsorption isotherm was of Langmuir.	62
21	Chill cast An alloy of Al-Zn-Mg	0.5M NaOH	*Hibiscus teterifa*	WL method was used. The process thermodynamics was also investigated.	63
22	Mild steel	1M HCl and H_2SO_4	Seeds and leaves extract of *Phyllanthus amarus*	WL and gasometric techniques were used. The Temkin isotherm fitted this process.	64
23	Mild steel	1M HCl and 0.5M H_2SO_4	*Murraya koenigii* leaves	Electrochemical measurements were done. The inhibitor was of mixed type. The adsorption isotherm was followed by Langmuir and Dubinin-Radushkevich isotherm.	65
24	Mild steel	1M H_2SO_4	Coriander, black cumin	WL, EIS and linear polarization were used.	66
25	Mild steel	2M H_2SO_4	Ethanolic extract of *Medicago sative*	The alternating current and direct current. electrochemical techniques were used.	67
26	Aluminium	0.5M NaOH	*Hibiscus sabdariffa* leaves	10% ethanol was used as additive. Hydrogen evolution followed by EIS, SEM, WL techniques were used.	68

Materials Research Forum LLC
https://doi.org/10.21741/9781644901496-1

27	X52 mild steel	20% H_2SO_4	*Cotula cinerae, Retama retam* plant	WL, Electrochemical methods were used.	69
28	Aluminium	Highly basic (pH=12)	*Hibiscus rosasinensis*	The additive was Zn^{2+}. The inhibitor was cathodic type. WL followed by AC impedance techniques were used.	70
29	Carbon steel, Nickel, Zinc	Neutral solution along with solutions containing acid as well as alkali	Lawsonia leaves	Polarization technique was used. IE order for carbon and nickel: alkaline<neutral<acid, whereas for zinc IE order: acid<alkaline<neutral	71
30	316 stainless steel	HCl (5%)	*Medicago polymorpha Roxb.*	The process was carried out at ambient temperature. Open circuit potential studies were done.	72
31	Aluminium, copper and brass	NaOH and acid chloride	Leaves of date palm and Lawsonia inermis, corn	WL and solution analysis were done. The adsorption isotherm was of Temkin isotherm.	73
32	N80 steel	15% HCl	Formaldehyde extract of different plant leaves	The synergistic effects were monitored. Thermodynamic parameters were calculated. The adsorption isotherm was of Langmuir isotherm.	74

33	An alloy of Al-2.5Mg	NaCl (3%)	Rosemary leaves extract containing phenol based compound	The analysis of curves of Potentiodynamic polarization was done. The adsorption process was followed by Freundlich isotherm. Ferulic acid was responsible for the inhibition action.	75
34	Aluminium	2M HCl	Mucilage extract of opuntia	Langmuir isotherm fitted the process. WL and hydrogen evolution techniques were used.	76
35	Aluminium	1M HCl	Roots of ginseng	WL method was followed. IE was 93.1%. The adsorption isotherm was Freundlich isotherm	77
36	Mild steel	3% NaCl	*Ricinus communis,* Coumarines plants	Galvanostatic anodic and cathodic polarization measurements were done.	78
37	Mild steel	0.2M HCl	Leaves and bark of mango plant	WL method was employed. 0.2M sulfuric acid concentration gave a good IE.	79
38	Mild steel	0.1M HCl	*Calotropis procera* plant extract	WL method was used and the polarization curves were analyzed.	80
39	Mild steel	20%, 50% and 88% H_3PO_4	*Zenthoxylum alatum* plant extract	WL, EIS, XPS and FTIR abalysus was carried out. The IE was 88%.	81
40	Tin	HNO_3	*Ficus carica L., Glycyrrhiza glabra* plants	Thermometric method was used.	82

Table 5 gives country wise information of the status of research on green corrosion inhibitors.

Table 5.   Green corrosion inhibitors currently in use (countrywise).

Name of country	Green inhibitors (from different plants)	Findings	Ref.
Nigeria and Jeddah	Aloe vera extract	Zinc in 2M HCl; aluminum in 0.5M HCl (84% IE)	83
Somalia and Italy	Seeds of Datura stramonium, Cassia occidentalis Papaia, and Poinciana pulcherrima and sap of Auforpio turkiale, Calotropis procera B, Azydracta indica and Papaia.	Mild steel in 1N HCl (88–96% IE)	84
China	Jasminum nudiflorum Lindl. leaves extract	Mixed type of inhibitor for cold rolled steel (CRS) in 1.0 M HCl	85
India	Vitis vinifera seed and skin extract	Pure copper and one of its alloy (Cu-27Zn brass) in a natural seawater environment (92.3% IE)	86
Columbia	Tobacco extracts	Aluminum and steel in acidic medium	87
Mexico	Opuntia-Ficus-Indica (Nopal) mucilage	Steel in a concrete like environment (alkaline solutions)	88
Latin America	Zygophyllum album	X52 mild steel in 1M H2SO4	89
Malaysia	Uncaria gambir extract	Mixed-type inhibitor for mild steel in aqueous (~70% IE)	90
Morocco, France	Rosmarinus officinalis oil	C38 steel in 2.0M H2SO4 (~60% IE)	91
Iran	Lawsonia inermis extract	Mixed type of inhibitor for mild steel in 1M HCl (92% IE)	92
French Guiana	Annona squamosa extract	Mixed-type inhibitor for C38 steel in 1M HCl	93

7.   Opportunities

Green corrosion inhibitors could be cost effective, i.e., economic feasibility is always a key factor. It is well known that many technological fields, households etc. are affected through many ways which in turn affect human beings. Several industries are running without corrosion prevention management. In 1978, National Bureau of Standards (USA) published a list of cost items which were affected by corrosion. The list includes industrial building, product loss, different kinds of parts of many plants and machines etc. [94]. The sacking of machines and minimization of production rate are greatly affected by these. A report published by National Association of Corrosion Engineers, worldwide corrosion authority in 2013 which shows that 2.5 trillion USD was spent worldwide due to corrosion which in turn equivalence to the worldwide gross domestic products as 3.4%. In practice of corrosion inhibition, about 35% i.e., 875 billion USD annually is saved on a global basis. This point gives a giant expansion of the use of green corrosion inhibitors. Many works show the use of bio-wastes and natural products as green corrosion inhibitors can maintain the economic feasibility. The inhibition of corrosion for metals and alloys such as different types of steel by using Phoenix clactylifera seeds, highlight the economic aspects of the bio-waste with the reference of a one seeded berry as fruit of the said plant [95]. The inhibition efficiency of the pericarp of the food is near about 97% at a temperature of 50°C. There is another patent which reported that some antimicrobial agents which is derived from plants have powerful tendency towards corrosion inhibition in the area of corrosion in pipelines for oil transportation, water-cooling systems, different kind of storage tanks, systems for fire protection etc. The 40% of these corrosion problems are greatly affected by microbes. [96]. Considering the fact of reduction of pollution, excluding the conventional protocols of the conventional methodologies there are some eco-friendly as well as biodegradable green corrosion inhibitors viz. sweet potato stems and stalks of lettuce flower stalks and stems of sweet potato, as reported in a Chinese patent [97].The importance of working on many wastes like seeds and skin, byproducts of mango, passion-fruit, fruit juices, cashew, and orange can play an important role in inhibition of corrosion as eco-friendly corrosion inhibitors. This study was performed by Gomes et al. [98] for prevention of steel corrosion in acid medium. Not only prevention of corrosion of steel but also of copper, aluminum and their individual alloys were investigated by another patent applying alternative formulations and method [99]. Dealing with corrosion inhibitors based on vegetable oils considering biorefinery and creating give-and-take relationship, more precisely synergistic relation between petroleum industries and natural resources and providing uniqueness in green methodologies. Versails worked and boosted the transformation of conventional sources towards the sustainable sites [100].

The field of corrosion inhibitors derived from naturally available products is scarcely applied to real-scale plants for arresting damage. The fault of using green products from vegetables, fruits or matrices as inhibitors is very much related with this aspect. These are precious as they are hard to harvest, scarce and edible. Irrespective of the source value the formulation of an extract from different natural substances can be dealt in the literature under lab-scale research. The probability of the natural extracts to be used in inhibition of corrosion is quite high, considering that the property of corrosion inhibition is closely related to unsaturation or the presence of hetero atoms in the source materials. The extracts from natural resources generally possess these kinds of features. Instead of the question of the impact of these inhibitors on society as well as economy, it is still very important considering the final application of the products.

The benefits of green corrosion inhibitors are tabulated in Table 6.

Table 6.   Benefits of green corrosion inhibitors.

Types of green corrosion inhibitors	% of efficiency	Cost	References
Plant leaves, latex, fruits	75-96%	Low	101
Plant stems, branches	87-95%	Low	102
Gum	34-46%	Low	103
Flour, yeast etc.	65-82%	Very low	104
Caffine, nicotine	80-90%	Medium	105
Tobacco	70-91%	Low	106
Vapour phase inhibitors	96%	Medium	107
Ionic liquids	79-99%	Expensive	108
Chitosan and proteins	98%	Medium	109
Leguminous seeds	34-91%	Low	110
Drugs	94-97%	Medium	111

8.   Challenges

Though the words "green corrosion inhibitors" are interesting, having lots of advantages relating to the environment, there are several limitations too which encourage the potentiality of taking the challenges while using these inhibitors. Considering green corrosion inhibitors, it must be noted whether the inhibitor is still non-toxic after a period of 28 days or not. Further the bioaccumulation of the inhibitor is also to be checked. Bioaccumulation is estimated by partition coefficient method between 1-octanol and water. Higher the partition coefficient higher will be the bioaccumulation. So, the components of the inhibitor must be pre-investigated before using it to prevent corrosion. One of the major drawbacks is their longevity. Since generally they are highly biodegradable it is difficult to store these for a long time. Choice of extraction method is another important fact. During the extraction of plants fragments it must be kept in mind that the solvent used for extraction must be less harmful. Also, some process requires long processing time, high

Materials Research Forum LLC
https://doi.org/10.21741/9781644901496-1

temperatures which disfavor the green corrosion inhibition. To overcome these problems nowadays supercritical fluid extraction process is applied as alternative techniques. But the effect of physico-chemical parameters on its mechanism is currently under investigation. Moreover, the active ingredients of the inhibitors are very much less. Nowadays some pharmaceuticals drugs are used for this purpose. But Gece pointed out that as all the drugs are not biodegradable it limits the processing of dry raw materials for green corrosion inhibitors. However, in spite of the above mentioned limitations, the green corrosion inhibitors are still the most promising alternatives because they are generally synthesized from natural products which in turn boost up their ready availability and cost effectiveness [112].

Conclusion

Metals - an essential part of our civilization - have the characteristic properties of degrading in presence of harmful atmospheric conditions created by oxygen or acidic, basic or even neutral conditions, which in turn compel us to think about preventing the phenomenon of corrosion. There are several techniques being applied to prevent corrosion. Among them use of inhibitor is the best option. Many countries have already used these green inhibitors to prevent corrosion. Due to environmental concern some bio-based inhibitors nowadays are well pronounced. These are also known as green corrosion inhibitors as they have no toxicity, are naturally available and very safe to handle. Moreover, the extraction process of the products is also generally cost-effective. As these products are easily available from economical point of view it is very feasible. From the literature survey it is inferred that the inhibition process is based on adsorption techniques which may follow Freundlich, Langmuir and Temkin isotherm processes. In a nutshell, it can be said that though green corrosion inhibitors have some limitations in terms of bioaccumulation, considering environmental protection these inhibitors are very much needed to make a cleaner, safer, sustainable surrounding and most importantly to get rid of the problems caused by corrosion.

References

[1] V.S. Sastri, Green Corrosion Inhibitor: Theory and Practice, A John Wiley & Sons Inc. Publication, 2011, pp.1. https://doi.org/10.1002/9781118015438

[2] Terminology Relating to Corrosion and Corrosion Testing, American Society for Testing and Materials Designation G 15-99b (Revised), 2000, 03.02.

[3] U.R. Evans, Metallic Corrosion and Protection, Edward Arnold, London, 1937

[4] C. Wagner, W. Traud, Uber die Deutung von Korrosionsvorg¨angendurch
¨Uberlagerung von elektrochemischenTeilvorgangen und uber die Potentialbildung an
Mischelektroden. Z. Elektrochem., 44 (1938) 391-402.

[5] D.M. Brasher, A.D. Mercer, Comparative study of factors influencing the action of
corrosion inhibitors for mild steel in neutral solution: I. sodium benzoate, Br. Corros.
J. 3(3) (1968) 120-129. https://doi.org/10.1179/000705968798326271

[6] A.K. Dunlop, R.L. Howard, P.J. Raifsnider, Mater. Prot. 8 (1969) 27.

[7] E. Ivanov, Y.I. Kuznetsov, Zasch. Met. 24 (1988) 36.

[8] P. Hersch, J.B. Hare, A. Robertson, An experimental survey of rust preventives in
water I. Methods of testing, J. Appl. Chem. 11 (1961) 246-250.
https://doi.org/10.1002/jctb.5010110704

[9] M.G. Fontana, R.W. Staehle, Advance Corrosion Science Technology, first volume,
Plenum Press, New York, London, 1970. https://doi.org/10.1007/978-1-4615-8252-6

[10] R. Char, T.L., D.K. Padma, Corrosion inhibitors for industry: Review of literature of
inhibitors minimizing corrosion in different environment from 1965-1968, Trans. Inst.
Chem. Eng. 47 (1969) 177-182.

[11] R. Walker, The use of benzotriazole as a inhibitor for copper, Anti-Corros. Meth.
Mater. 17(9) (1970) 9-15. https://doi.org/10.1108/eb006791

[12] V.S. Sastri, Green Corrosion Inhibitor: Theory and Practice, A John Wiley & Sons
Inc., Publication, 2011, pp. 258. https://doi.org/10.1002/9781118015438

[13] C. Xu, W. Gao, Pilling-Bedworth ratio for oxidation of alloys. Mater. Res. Innov. 4
(2000) 231-235. https://doi.org/10.1007/s100190050008

[14] R.E. Bedworth, N.B. Pilling, The oxidation of metals at high temperatures. J. Inst.
Metals. 29 (1923) 529-582.

[15] Omnia S. Shehata, Lobna A. Korshed, Adel Attia, Green Corrosion Inhibitors, Past,
Present, and Future, Chapter 6, IntechOpen, 2017.
https://doi.org/10.5772/intechopen.72753

[16] K.M. Emran, S.M. Ali, H.A. Al Lehaibi, Green methods for corrosion control,
Chapter 3, in: M. Aliofkhazraei (Ed.), Corrosion Inhibitors, Principles and Recent
Applications, IntechOpen, 2017. https://doi.org/10.5772/intechopen.72762

[17] S.K. Sharma (Ed.), Green Corrosion Chemistry and Engineering, Wiley-VCH
Verlag & Co. KGaA, 2012.

[18] I.H. Farooqi, M.A. Quraishi, Proceedings of the GLOCORR'2002, International Congress on Emerging Corrosion Control Strategies for the new Millennium, NCCI & CECRI, New Delhi, February 2002.

[19] V.S. Agarwaala, A new approach in corrosion inhibition, Proceedings of International Congress on Metallic Corrosion, Toronto, 1 (1984) 80.

[20] R.M. Saleh, A.A. Ismail, A.H. El Hosary, Corrosion inhibition by naturally occurring substances: The effect of aqueous extracts of some leaves and fruit peels on the corrosion of steel, Al, Zn and Cu in acids, Br. Corros. J. 17 (1982) 131-135. https://doi.org/10.1179/000705982798274345

[21] M. Abdel-Gaber, B.A. Abd-El-Nabey, I.M. Sidahmed, A.M. El-Zayaday, M. Saa dawy, Inhibitive action of some plant extracts on the corrosion of steel in acidic media, Corros. Sci. 48 (2006) 2765-2779. https://doi.org/10.1016/j.corsci.2005.09.017

[22] A. El-Hosary, R.M. Saleh, A.M. Sharns El Din, Corrosion inhibition by naturally occurringsubstances - I. The effect of Hibiscus subdariffa (karkade) extract on the dissolutionof Al and Zn, Corros. Sci. 12(1972) 897-904. https://doi.org/10.1016/S0010-938X(72)80098-2

[23] R. Ananda, L. Sathiyanathan, S. Maruthamuthu, M. Selvanayagam, S. Mohanan, N. Palaniswamy, Corrosion inhibition of mild steel by ethanolic extracts of Ricinus communis leaves, Indian J. Chem. Technol. 12 (2005) 356-360.

[24] M. Kliskic, J. Radosevic, S. Gudic, V. Katalinic, Aqueous extract of Rosmarinus officinalis L. as inhibitor of Al-Mg alloy corrosion in chloride solution, J. Appl. Electrochem. 30 (2000) 823-830. https://doi.org/10.1023/A:1004041530105

[25] B. Muller, Corrosion inhibition of aluminium and zinc pigments by saccharides, Corros. Sci. 44 (2002) 1583-1591. https://doi.org/10.1016/S0010-938X(01)00170-6

[26] F. Zucchi, I. Omar, Plant extracts as corrosion inhibitors of mild steel in HCl solutions, Surf. Tech. 24 (1985) 391-399. https://doi.org/10.1016/0376-4583(85)90057-3

[27] A. Minhaj, P.A. Saini, M.A. Quarishi, I.H. Farooqi, A study of natural compounds as corrosion inhibitors for industrial cooling systems, Corros. Prev. Control. 46 (1999) 32-38.

[28] A.Y. El-Etre, Natural honey as corrosion inhibitor for metals and alloys of copper in neutral aqueous solution, Corros. Sci. 40 (1998) 1845-1850. https://doi.org/10.1016/S0010-938X(98)00082-1

[29] M.J. Sanghvi, S.K. Shukla, A.N. Misra, M.R. Mehta, 5th National Congress on Corrosion Control, New Delhi, 1995, pp. 46.

[30] E. Oguzie, Inhibition of acid corrosion of mild steel by Telfaria occidentalis, Pigm. Resin Technol. 34 (2005) 321-326. https://doi.org/10.1108/03699420510630336

[31] L.R. Chauhan, G. Gunasekaran, Corrosion inhibition of mild steel by plant extract in dilute HCl medium, Corros. Sci. 49 (2007) 1143-1161. https://doi.org/10.1016/j.corsci.2006.08.012

[32] Y. Li, P. Zhao, Q. Liang, B. Hou, Berberine as a natural source inhibitor for mild steel in 1M H2SO4. Appl. Surf. Sci., 252 (5) (2005) 1245-1253. https://doi.org/10.1016/j.apsusc.2005.02.094

[33] G. Tammann, The chemical and galvanic properties of alloy states and their atomic configurations, Z. Anorg. Chem. 107 (1919) 155-156.

[34] C. Okafor, M.E. Ikpi, I.E. Uwah, E.E. Ebenso, U.J. Ekpe, S.A. Umoren, Inhibitory action of Phyllanthusamarus extracts on the corrosion of mild steel in acidic media, Corros. Sci. 50 (2008) 2310-2317. https://doi.org/10.1016/j.corsci.2008.05.009

[35] N. Eddy, Inhibitive and adsorption properties of ethanol extract of Colocasia esculenta leaves for the corrosion of mild steel in H2SO4, Int. J. Phys. Sci. 4 (2009) 165-171. https://doi.org/10.1108/03699421011085849

[36] S. Cheng, S. Chen, T. Liu, X. Chang, Y. Yin, Carboxymethyl chitosan as an eco-friendly inhibitor for mild steel in 1M HCl, Mater. Lett. 61 (2007) 3276-3280. https://doi.org/10.1016/j.matlet.2006.11.102

[37] O. Ime, O.E. Nelson., An interesting and efficient green corrosion inhibitor for Aluminium from extracts of Chlomolaena odorata L. in acidic solution. J. Appl. Electrochem. 40 (2010) 1977-1984 https://doi.org/10.1007/s10800-010-0175-x

[38] P.D. Rani, S. Selvaraj, Inhibitive and adsorption properties of punica granatum extract on brass in acid media, J. Phytology 2 (2010) 58-64.

[39] E. Khamis, N. Alandis, Herbs as new type of green inhibitors for acidic corrosion of steel. Material Wissenschaft und Werkstofftechnik. 33(2002) 550-554. https://doi.org/10.1002/1521-4052(200209)33:9<550::AID-MAWE550>3.0.CO;2-G

[40] M. Kundu, S.K. Prasad, K. Virendra, A review article on green inhibitors of reinforcement concrete corrosion. Int. J. Emerg. Res. Mangt. Technol. 5 (2016) 42-46.

[41] S.A. Asipita, M. Ismail, M.Z.A. Majid, C.S. Abdullah, J. Mirza, Green Bambusa arundinacea leaves extract as a sustainable corrosion inhibitor in steel reinforced

concrete. J. Cleaner Prod. 67 (2014) 139-146.
https://doi.org/10.1016/j.jclepro.2013.12.033

[42] D.G. Eyu, H. Esah, C. Chukwuekezie, J. Idris, I. Mohammad, Effect of green inhibitor on the corrosion behavior of reinforced carbon steel in concrete. ARON J. Eng. Appl. Sci. 8 (2013) 326-332.

[43] A.S. Fouda, A.H. Badr, Aqueous extract of propolis as corrosion inhibitor for carbon steel in aqueous solutions, Afr. J. Pure Appl. Chem. 7 (2013) 50-359. https://doi.org/10.5897/AJPAC2013.0524

[44] M.M. Fares, A.K. Maayta, M.M. Al-Qudah, Pectin as promising green corrosion inhibitor of aluminum in hydrochloric acid solution, Corros. Sci. 60 (2012) 112-117. https://doi.org/10.1016/j.corsci.2012.04.002

[45] N. Saidi, H. Elmsellem, M. Ramdani, A. Chetouani, K. Azzaoui, F. Yous, A. Aouniti, B. Hammouti, Using pectin extract as eco-friendly inhibitor for steel corrosion in 1 M HCl media, Der. Pharma. Chem. 7 (2015) 87-94.

[46] A.K. Larios-Galvez, J. Porcayo-Calderon, V.M. Salinas-Bravo, J.G. Chacon-Nava, J.G. Gonzalez-Rodriguez, L. Martinez-Gomez, Use of Salvia hispanica as an eco-friendly corrosion inhibitor for bronze in acid rain, Anti-Corros. Methods Mater. 64 (2017) 654-663. https://doi.org/10.1108/ACMM-02-2017-1760

[47] A. Ostovari, S.M. Hoseinieh, M. Peikari, S.R. Shadizadeh, S.J. Hashemi, Corrosion inhibition of mild steel in 1 MHCl solution by henna extract: A comparative study of the inhibition by henna and its constituents (Lawsone, Gallic acid, α-d-Glucose and Tannic acid), Corros. Sci. 51 (2009) 1935-1949. https://doi.org/10.1016/j.corsci.2009.05.024

[48] D. Amar, S. Lahcene, D. Djamila, O. Meriem, N. Abdelkader, R. Salah-edin, Alkaloids Extract from Peganumharmala Plant as Corrosion Inhibitor of 6063 Aluminium Alloy in 1 M Hydrochloric Acid Medium, J. Chem. Pharm. Res. 9 (2017) 311-318.

[49] H. Gerengi, H.I. Sahin, Schinopsis lorentzii extract as a green corrosion inhibitor for low carbon steel in 1 M HCl solution, Ind. Eng. Chem. Res. 51 (2012) 780-787. https://doi.org/10.1021/ie201776q

[50] M. Faustin, A. Maciuk, P. Salvin, C. Roos, M. Lebrini, Corrosion inhibition of C38 steel by alkaloids extract of Geissospermumlaeve in 1M hydrochloric acid: Electrochemical and phytochemical studies, Corros. Sci. 92 (2015) 287-300. https://doi.org/10.1016/j.corsci.2014.12.005

[51] R. Fdil, M. Tourabi, S. Derhali, A. Mouzdahir, K. Sraidi, C. Jama, A. Zarrouk, F. Bentiss, Evaluation of alkaloids extract of Retamamonosperma (L.) Boiss. stems as a green corrosion inhibitor for carbon steel in pickling acidic medium by means of gravimetric, AC impedance and surface studies, J. Mater. Environ. Sci. 9 (2018) 358-369. https://doi.org/10.26872/jmes.2018.9.1.39

[52] P.B. Raja, A.K. Qureshi, A.A. Rahim, H. Osman, K. Awang, Neolamarckiacadamba alkaloids as eco-friendly corrosion inhibitors for mild steel in 1 M HCl media, Corros. Sci. 69 (2013) 292-301. https://doi.org/10.1016/j.corsci.2012.11.042

[53] A. Y. El-Etre, Inhibition of acid corrosion of carbon steel using aqueous extract of olive leaves, J. Colloid Interface Sci. 314 (2007) 578-583. https://doi.org/10.1016/j.jcis.2007.05.077

[54] N.A. Odewunmi, S.A. Umoren, Z.M. Gasem, Watermelon waste products as green corrosion inhibitors for mild steel in HCl solution, J. Environ. Chem. Eng. 3(2015) 286-296. https://doi.org/10.1016/j.jece.2014.10.014

[55] A.N. Grassino, J. Halambek, S. Djakovi'c, S. Rimac Brnci'c, M. Dent, Z. Grabari'c, Utilization of tomato peel waste from canning factory as a potential source for pectin production and application as tin corrosion inhibitor, Food Hydrocoll. 52 (2016) 265-274. https://doi.org/10.1016/j.foodhyd.2015.06.020

[56] A. Bouyanzer, B. Hammouti, L. Majidi, Pennyroyal oil from Menthapulegium as corrosion inhibitor for steel in M HCl, Materials Letters. 60(2006)2840-2843 https://doi.org/10.1016/j.matlet.2006.01.103

[57] K. Boumhara H. Harhar, M. Tabyaoui A. Bellaouchou A. Guenbour A. Zarrouk, Corrosion inhibition of mild steel in 0.5 m H2SO4 solution by artemisia herba-alba Oil, J. bio- tribo-corros. 8 (2019) Article 5. https://doi.org/10.1007/s40735-018-0202-8

[58] Narayanasamy, Poongothai, P. Rajendran, M. Natesan, N. Palaniswamy, Wood bark oils as vapour phase corrosion inhibitors for metals in NaCl and SO2 environments. Ind. J. Chem. Technol. 12 (2005) 641-647.

[59] M.E.I-Sayed, O.Y. Mansour, I.Z. Selim, M.M. Ibrahim, Identification and utilization of banana plant juice and its liquor as anti-corrosive mate-rials, J. Sci. Ind. Res. 60 (2001) 738-747.

[60] A.K. Satapathy, G. Gunasekaran, S.C. Sahoo, K. Amit, P.V. Rodrigues, Corrosion inhibition by Justiciagendarussa plant extract in hydrochloric acid solution, Corros. Sci. 51 (2009) 2848-2856. https://doi.org/10.1016/j.corsci.2009.08.016

[61] F.S.de Souza, A. Spinelli, Caffeic acid as a green corrosion inhibitor in mild steel, Corros. Sci. 51 (2009) 642-649. https://doi.org/10.1016/j.corsci.2008.12.013

[62] J.C. Rocha, J.A. Cunha, P. Gomes, E.D. 'Elia, Corrosion inhibition of carbon steel in hydrochloric acid solution by fruit peel aqueous extract, Corros. Sci. 52 (2010) 2341-2348 https://doi.org/10.1016/j.corsci.2010.03.033

[63] F.A. Ayeni, V.S. Aigbodion, S.A. Yaro, Non-toxic plant extract as corrosion inhibitor for chill cast Al-Zn-Mg alloy in caustic soda solution, EurAsia Chemico-Technol. J. 9 (2007) 91-96.

[64] P.C. Okafor, M.E. Ikpi, I.E. Uwah, E.E. Ebenso, U.J. Ekpe, S.A. Umoren, Inhibitory action of Phyllanthusamarus extracts on the corrosion of mild steel in acidic media, Corros. Sci. 50 (2008) 2310-2317. https://doi.org/10.1016/j.corsci.2008.05.009

[65] M.A. Quraishi, A. Singh, V.K. Singh, D.K. Yadav, A.K. Singh, Green approach to corrosion inhibition of mild steel in hydrochloric acid and sulphuric acid solutions by the extract of Murrayakoenigii leaves, Mater. Chem. Phys. 122 (2010) 114-122. https://doi.org/10.1016/j.matchemphys.2010.02.066

[66] E. Khamis, N. Alandis, Herbs as new type of green inhibitors for acidic corrosion of steel, Materialwissenschaftund Werkstofftechnik. 33 (2002) 550-554 https://doi.org/10.1002/1521-4052(200209)33:9<550::AID-MAWE550>3.0.CO;2-G

[67] A.M. Al-Turkustani, S.T. Arab, L.S.S. Al-Qarni, MedicagoSative plant as safe inhibitor on the corrosion of steel in 2.0 M H2SO4 solution, J. Saudi Chem. Soc. 15(1) (2011) 73-82. https://doi.org/10.1016/j.jscs.2010.10.008

[68] E.A. Noor, Potential of aqueous extract of Hibiscus sabdariffa leaves for inhibiting the corrosion of aluminum in alkaline solutions, J. Appl. Electrochem. 39 (2009) 1465-1475. https://doi.org/10.1007/s10800-009-9826-1

[69] M. Dakmouche, S. Ladjel, N. Gherraf, M. Saidi, M. Hadjaj, M.R. Ouahrani, Inhibition effect of some plant extracts on the corrosion of mild steel in H2SO4 medium, Asian J. Chem. 21 (2009) 6176-6180.

[70] S. Rajenderan, J. Jeyasundari, P. Usha, J.A. Selvi, B. Narayanasamy, A.P.P. Regis, P. Renga, Corrosion behavior of aluminium in the presence of an aqueous extract of hibiscus rosasinensis, Port. Electrochim. Acta. 27 (2009) 153-164. https://doi.org/10.4152/pea.200902153

[71] A.Y. El-Etre, M. Abdallah, Z.E. El-Tantawy, Corrosion inhibition of some metals using lawsonia extract, Corros. Sci. 47 (2005) 385-395. https://doi.org/10.1016/j.corsci.2004.06.006

[72] T.K. Soror, New naturally occurring product extract as corrosion inhibitor for 316 stainless steel in 5% HCl, J. Mater. Sci. Technol. 20 (2004) 463-466.

[73] H.H. Rehan, Corrosion control by water-soluble extracts from leaves of economic plants, Material wissenschaft und Werkstofftechnik. 34 (2003) 232-237. https://doi.org/10.1002/mawe.200390034

[74] T. Emranuzzaman, S. Kumar, G. Vishwanatham, Udayabhanu, Synergistic effects of formaldehyde and alcoholic extract of plant leaves for protection of N80 steel in 15% HCl, Corros. Eng. Sci. Technol. 39 (2004) 327-332. https://doi.org/10.1179/174327804X13181

[75] J. Radosevic, M. Kliskic, A. Visekruna, Inhibition of corrosion of the Al-2.5Mg Alloy by means of the third acidic phenolic subfraction of aqueous extract of Rosemary, J. Chem. Chem. Eng. 50 (2001) 537-541.

[76] A.Y. El-Etre, Inhibition of aluminum corrosion using Opuntia extract, Corros. Sci. 45 (2003) 2485-2495. https://doi.org/10.1016/S0010-938X(03)00066-0

[77] I.B. Obot, N.O. Obi-Egbedi, Ginseng, a new efficient and effective eco-friendly corrosion inhibitor for aluminium alloy of type AA 1060 in hydrochloric acid solution, Int. J. Electrochem. Sci. 4 (2009) 1277-1288.

[78] Y.F. Barakat, A.M. Hassan, A.M Baraka, Corrosion inhibition of mild steel in aqueous solution containing H2S by some naturally occurring substances, Materialwissenchaft und Werkstofftechnik. 29 (1998) 365-370. https://doi.org/10.1002/mawe.19980290709

[79] C.A. Loto, The effect of mango bark and leaf extract solution additives on the corrosion inhibition of mild steel in dilute sulphuric acid-part I, Corros. Prev. Control. 48 (2001) 38-41.

[80] G.H. Awad, Effect of some plant extracts on the corrosion of mild steel in 0.1N Hydrochloric acid solutions., 1985.

[81] G. Gunasekaran, L.R. Chauhan, Eco-friendly inhibitor for corrosion inhibition of mild steel in phosphoric acid medium, Electrochim. Acta. 49(25) (2004) 4387-4395. https://doi.org/10.1016/j.electacta.2004.04.030

[82] M.E. Ibrahim, A.M. El-Khrisy, E.M.M. Al-Abdallah, A.Baraka, evaluation of the inhibitor action of certain natural substances used as corrosion inhibitors-1 in the dissolution of tin in nitric acid, Mo MetalloberflancheBeschichten won Metall und Kunstsoff, (1981) 134-136.

Materials Research Foundations**107** (2021) 1-29   https://doi.org/10.21741/9781644901496-1

[83] A. Olusegun, K. and James, A.O. (2010) Corros. Sci., 52 (2), 661-664 https://doi.org/10.1016/j.corsci.2009.10.026

[84] Z. Fabrizio, O.I. Hashi, Plant extracts as corrosion inhibitors of mild steel in HCl solutions, Surf. Technol. 24(4) (1985) 391-399. https://doi.org/10.1016/0376-4583(85)90057-3

[85] X.-H. Li, S.-D. Deng, H. Fu, Inhibition by Jasminum nudiflorum Lindl. leaves extract of the corrosion of cold rolled steel in hydrochloric acid solution, J. Appl. Electrochem., 40(9) (2010) 1641-1649. https://doi.org/10.1007/s10800-010-0151-5

[86] P.D. Rani, S. Selvaraj, Inhibitive action of vitis vinifera (grape) on copper and brass in natural sea water environment, Rasayan J. Chem., 3(3) (2010) 473-482.

[87] G.D. Davis, J.A. von Fraunhofer, L.A. Krebs, C.M. Dacres, Corros. 2001, Paper 1558.

[88] A.A. Torres-Acosta, Opuntia-Ficus-Indica (Nopal) mucilage as a steel corrosion inhibitor in alkaline media, J. Appl. Electrochem. 37(7) (2007) 835-841. https://doi.org/10.1007/s10800-007-9319-z

[89] N. Gherraf, T.Y. Namoussa, S. Ladjel, M.R. Ouahrani, R. Salhi, A. Belmnine S. Hameurlain, B. Labed, Am.-EurAsian J. Sustain. Agric., 3 (4) (2008) 781-783.

[90] H.M. Hazwan, K.M. Jain, Electrochemical studies of mild steel corrosion inhibition in aqueous solution by uncaria gambir extract, J. Phys. Sci. 21 (1) (2010) 1-13.

[91] E.E. Ouariachi, J. Paolini, M. Bouklah, A. Elidrissi, A. Bouyanzer, B. Hammouti, J.-M. Desjobert, J. Costa, Acta Metall. Sin. (Engl. Lett.) 23 (1) (2010) 13-20.

[92] A. Ostovari, S.M. Hoseinieh, M. Peikari, S.R. Shadizadeh, S.J. Hashemi, Corrosion inhibition of mild steel in 1 M HCl solution by henna extract: A comparative study of the inhibition by henna and its constituents (Lawsone, Gallic acid, α-D-Glucose and Tannic acid), Corros. Sci. 51 (9) (2009) 1935-1949 https://doi.org/10.1016/j.corsci.2009.05.024

[93] M. Lebrini, F. Robert, C. Roos, Inhibition effect of alkaloids extract from annona squamosa plant on the corrosion of c38 steel in normal hydrochloric acid medium, Int. J. Electrochem. Sci. 5 (2010) 1698-1712.

[94] G. Bianchi, F. Mazza, Corrosione e ProtezionedeiMetalli; AssociazioneItaliana di Metallurgia: Milano, Italy, 2000.

[95] G. Koch, J. Varney, N. Thompson, O. Moghissi, M. Gould, J. Payer, International Measures of Prevention, Application, and Economics of Corrosion Technologies Study; NACE Internationa: Houston, TX, USA, 2016.

[96] G.R. Thusnavis, K.P.V. Kumar, Green corrosion inhibitor for steel in acid medium, Application No. 6278/CHE/2014 A, 12 December 2014.

[97] J.B. Zhang, F. Zhenquan, Y. Yaorong, Z. Chengxian, L. Xuehui, C.R. Fagen, Extract Corrosion Inhibitor of Sweet Potato Stems and Lettuce Flower Stalks and Preparation Method Thereof. Patent No. CN102492948B, 31 July 2013.

[98] J.A. Ponciano Gomes, J.R. Cardoso, E. D'Elia, Use of Fruit Skin Extracts as Corrosion Inhibitors and Process for Producing Same. U.S. Patent US8926867B2, 6 January 2015.

[99] P.J. Kinlen, L.P.S. Pinheiro, Methods and apparatuses for selecting natural product corrosion inhibitors for application to substrates. U.S. Patent Application No. US 2018/0202051 Al, 19 July 2018.

[100] R. Lima, A. Casalini, A. Palumbo, G. Rivici, Corrosion inhibitor comprising complex oligomeric structures derived from vegetable oils. Patent Application No. WO 2017/140836 Al, 24 August 2017.

[101] E.I. Ating, S.A. Umoren, I.I. Udousoro, E.E. Ebenso, A.P. Udoh, Leaves extract of Ananassativum as green corrosion inhibitor for aluminum in hydrochloric acid solution. Green Chem. Lett. Rev. 3 (2010) 61-68. https://doi.org/10.1080/17518250903505253

[102] E. Honarmand, H. Mostaanzadeh, M.H. Motaghedifard, M. Hadia, M. Khayadkashanic, Inhibition effect of opuntia stem extract on corrosion of mild steel: a quantum computational assisted electrochemical study to determine the most effective components in inhibition. Protect Met. Phys. Chem. Surface, 53 (2017) 560-572. https://doi.org/10.1134/S207020511703008X

[103] M. Abdallah, Guar gum as corrosion inhibitor for carbon steel in sulfuric acid solutions. Port. Electrochim. Acta. 22 (2004) 161-175. https://doi.org/10.4152/pea.200402161

[104] Baldwin J. British Patent. 1895: 2327.

[105] A. Espinoza-Vázquez, F.J. Rodríguez-Gómez, Caffeine and nicotine in 3% NaCl solution with CO2 as corrosion inhibitors for low carbon, RSC Adv. 74 (2016) 70226-70236. https://doi.org/10.1039/C6RA07673D

[106] C.A. Loto, A.P.I. Popoola, Effect of tobacco and kola tree extracts on the corrosion inhibition of mild steel in acid chloride. Int. J. Electrochem. Sci. 6 (2011) 3264-3276.

[107] H. Kumar, V. Saini, V. Yadav, Study of vapour phase corrosion inhibitors for mild steel under different atmospheric conditions. Int. J. Eng. Innov. Technol. 3 (2013) 206-211.

[108] S.K. Shetty, A.N. Shetty, Ionic liquid as an effective corrosion inhibitor on 6061 Al-15 Vol. Pct. SiC (p) composite in 0.1 M H2SO4 medium-an ecofriendly approach. Can. Chem. Trans. 3 (2015) 41-64. https://doi.org/10.13179/canchemtrans.2015.03.01.0160

[109] R. Oukhrib, B. El Ibrahim, H. Bourzi, K. El Mouaden, A. Jmiai, S. El Issami, L. Bammou, L. Bazzi, Quantum chemical calculations and corrosion inhibition efficiency of biopolymer "chitosan" on copper surface in 3% NaCl. J. Mater. Environ. Sci. 8 (2017) 195-208.

[110] Subhashini S. Corrosion inhibition study of mild steel in acid media by extract of some leguminous seeds as eco-friendly inhibitors. J. Camel Pract. Res. 11 (2004) 27-34.

[111] C. Verma, D.S. Chauhan, M.A. Quraishi, Drugs as environmentally benign corrosion inhibitors for ferrous and nonferrous materials in acid environment: an overview. J. Mater. Environ. Sci. 8 (2017) 4040-4051

[112] R.M. Palou, O. Olivares-Xomelt, N.V. Likhanova, Environmentally friendly corrosion inhibitors, in: Developments in Corrosion Protection, M. Aliofkhazraei (Ed.), InTech, 2014, pp. 431-465.

Materials Research Forum LLC
https://doi.org/10.21741/9781644901496-2

Chapter 2

Sources of Active Ingredients for Sustainable Corrosion Inhibitors

Rekha Sharma[1], Sapna Nehra[2], and Dinesh Kumar[3*]

[1]Department of Chemistry, Banasthali University, Rajasthan-304022, India

[2]Department of Chemistry, Dr. K. N. Modi University, Newai, Rajasthan 304021, India

[3]School of Chemical Sciences, Central University of Gujarat, Gandhinagar-382030, India

* dinesh.kumar@cug.ac.in

Abstract

Metal corrosion is a grave problem, having deleterious effects on human health, the economy, the environment, and many engineering schemes, for example, automobiles, aircraft, naval vessels, and pipelines. For the confirmation of enduring reliability and stability of alloys and metals, corrosion-protective surfaces are of the greatest significance, for example, ions and water, through restraining their interactions with corrosive species. Though, their applied submissions are frequently bounded whichever through deprived mechanical robustness or else through the incapability to resist low surface tension liquids, for example, alcohols and oil. In this chapter, we have focused on diverse materials as sustainable corrosion inhibitors such as organic corrosion inhibitors, green corrosion inhibitors, and polymer-based corrosion inhibitors to protect materials from being corroded. Amongst them, we especially focus on green corrosion inhibitors as a consequence of simple manufacturing, easy availability, cost-effectiveness, and biodegradable nature.

Keywords

Cost-Effective, Sustainable, Renewable, Corrosion, Applications

Contents

1. Introduction

Corrosion is an enhanced method of aging that producing hefty commercial forfeitures at the international/national scale and takes place through electrochemical or else chemical reactions of resources with their environments [1]. Corrosion is a thermosensitive method that can be condensed to a certain degree through using many methods such as designing, surface cleaning, use of inhibitors, alloying, coatings, and paints, but it cannot be eliminated [2−5]. Because of their simple submission and periodic shielding of industrial infrastructures, synthetic polymer coatings have been widely utilized amongst these methods to avert corrosion, by which it can reduce the cost of corrosion and its conservation [6]. Though, for the growth of maintainable polymers, via adaptable monomers, the impression of emission of VOCs and fossil fuel depletion through their dispensation have fascinated the researcher's attention [7]. Out of them, low molecular weight containing vegetable oil-based polymers of having decent variability with insignificant or no carbon-based diluents and invent common submissions in flavors, pastes, coatings, and dyes [8,9].

2. Corrosion inhibitors

In various corrosive media, to discover corrosion inhibitors of organic origin, substantial exertions have been organized over the years [10-13]. For protecting various materials from corrosion various compounds such as sulfur-containing compounds, N_2 based materials in an acidic medium and their byproducts, acetylenic, thioaldehydes complexes, aldehydes, and numerous alkaloids, for instance, strychnine, quinine, nicotine, and papaverine are utilized as corrosion inhibitors, whereas, nitrite, benzoate, phosphate, and chromate perform as decent inhibitors in neutral media. Inhibitors avert or diminution of the metal reacts with the reaction medium. These inhibitors decrease the degree of corrosion through the following methods:

- intended for substrates to the surface of metal, plummeting the degree of diffusion

- onto the surface of the metal, sorption of molecules/ions

- decreasing or increasing the cathodic or anodic reaction,

- inhibitors that have in situ submission benefit and are frequently easy to use

Many features such as easy obtainability, amount and cost, and vitally wellbeing to the atmosphere and its species are required for consideration when selecting an inhibitor.

2.1   Organic inhibitors

Organic inhibitors have a higher electron density and basicity and therefore perform as a corrosion inhibitor, which usually partakes heteroatoms such as N, S, and O. On the metal surface, these heteroatoms (N, O, and S) are the lively cores' adsorption. The sequence O < N < S < P ought to follow for inhibition competence. To decrease the occurrence of steel deterioration, the usage of organic composites comprising O, S, and specifically N has been studied earlier. The outcomes displayed that on the metal surface, utmost organic inhibitors adsorbed through forming a compact barrier with water molecules and moving it on the surface. The electron transfers to the metal from the inhibitor facilitated through the accessibility of p-electrons and lone pair (nonbonded) in inhibitor molecules. A coordinate covalent bond can be shaped concerning the allocation of electrons to the surface of the metal ion from the inhibitor. To recover the inhibition process various functional groups i.e., Nitro, Aldehyde, Carboxylic acid, or Amine is substituting the C in the ring bonded to an H atom [13]. The electron density of metal varies at the site of addon ensuing in the deceleration of the anodic or cathodic responses. Straight chain amines have been observed containing carbons between 3-14. In the capability to prevent corrosion arises, inhibition small decrease or increase with higher members, nonetheless sharply increases with 10 carbons number in the carbon chain. Though, the solubility of the inhibitors may be increased due to the attendance of hydrophilic functional moieties. The physicochemical possessions and chemical structure of the composite are the outcomes of organic inhibitor, for example p-orbital character, functional moieties, molecular electronic assembly, and the electronic mass at the contributor particle. The inhibition might be possible because of the following points:

- Increase in an overvoltage of cathodic and anodic site

- Adsorption of molecules or ions of cathodic or anodic places

- The establishment of a protecting fence film.

The aspects that contribute to inhibitory action are:

Sustainable Corrosion Inhibitors   Materials Research Forum LLC
Materials Research Foundations107 (2021) 30-45   https://doi.org/10.21741/9781644901496-2

- Molecular size
- Bond strength and chain length
- Atomic bonding
- Atmospheric solubility
- Cross-linking ability.

In contradiction of acid attack, to create a fence of solitary or numerous molecular layers, the inhibitors play an important role. Concerning the charge transfer beginning from one to another end, they frequently relate this protecting act to physical or chemical adsorption. Having several substituents, S and N, comprising heterocyclic complexes, are measured to be an operative deterioration inhibitor. In acid solutions, to inhibit corrosion of metals, some materials propose a special affinity, for example, hydrazine derivatives and thiophene. Inorganic materials, for example, chromates, phosphates, silicates, dichromates, tungstates, borates, arsenates, and molybdates, have been originating operative as metal deterioration inhibitors. In acid conditions, spinoffs of pyrrole are supposed to display a decent shield contrary to deterioration. In preparing anticorrosive and primers coatings, these inhibitors have also attained helpful submission. For the adsorption process, the inhibitive features of such composites acting as the reaction center. To protect metal disbanding from acid solutions, inhibitors are frequently added in manufacturing procedures. To progress nontoxic, inexpensive, and ecological procedures have fascinated the researcher's attention to the emphasis on the usage of natural products, the recognized dangerous belongings of utmost artificial organic inhibitors. Progressively, considering Cr(VI) existence debarred and categorized as a hazard, it is required to establish cultured novel coverings for the enhanced recital. The utmost choice of shielding metals contrary to corrosion is the usage of inhibitors. Many inhibitors in usage are selected thru composites partaking heteroatoms in their long-chain or aromatic carbon system, and synthesized by inexpensive raw material, out of them, most of these inhibitors are poisonous to the atmosphere. This fascinated the researcher's attention to develop corrosion inhibitors using green method [14].

3.   Green inhibitors

These inhibitors do not comprise toxic compounds such as heavy metals and other elements, and these are biodegradable. In alkaline or acidic atmosphere, to hinder the metal corrosion, an efficacious usage of naturally ensuing constituents has been reported by some research groups.

As a deterioration inhibitor for Cu, El-Etre reported natural honey [15], and also reported opuntia extract on aluminum as an inhibitor [16], at 25 °C for the Al+2.5Mg alloy, rosemary leaves described as deterioration inhibitor in a 3% NaCl solution [17], and the deterioration of aluminum in HCl inhibited by Delonix regia extracts [18]. In HCl solution, the deterioration of SX 316 steel was regulated via potentiostatic technique in addition to weight loss capacities using the seed extract of Ammi visnaga (khillah). Because of interaction among iron cations, the method of action is accredited to form insoluble compounds [19]. The ethanolic extract of Carica papaya leaves were utilized for the corrosion inhibition of mild steel in H_2SO_4 by Ebenso et al. [20]; Corrosion inhibition of mild steel by African bush pepper (Piper guinensis) [21]; Corrosion inhibition in H_2SO_4 of mild steel using Azadirachta indica leaves extract (neem) [22]. Zucchi and Omar [23] reported extracts of different plant i.e., Poinciana pulcherrima, Papaia, Datura stramonium seeds, and Cassia occidentalis, and Auforpio turkiale, Azadirachta indica, Calotropis Procera B, and Papaya for their potential towards corrosion inhibition. The outcomes displayed that the steel corrosion reduced by all the plant extracts excluding Azadirachta indica and Auforpio turkiale with efficiency in 1N HCl of 88%–96% and which decreases slightly in 2N HCl. Umoren et al. [24], reported the mild steel corrosion inhibition in H_2SO_4 in the attendance of synthetic polymer i.e., polyethylene glycol (PEG) and naturally occurring polymer i.e., gum arabic (GA), they attributed the outcome of these plants to the harvests of protein content hydrolysis [23]. The results showed that GA was found to be less effective than that of PEG.

Particularly in environmental technologies and sustainable energy, Gong et al., 2016 reported graphene with pristine graphene and its analogs of GO and RGO. There are many effective belongings of graphene submissions, from detecting energy storage and conversions to ecological treatment. Graphene is virtually enduring united by alternative vigorous substantial as a compound in preference to the occurrence as a separate operating substantial. The interaction through graphene benefits to reduce such limits in the unadorned used material and emphasizes them with its effective transfer of heat, elevated surface to bulk ratio, and conduction of an electron. The approach of exterior functionalization of graphene to accomplish this grasps the significant to permitting the production of high-performance complexes, with either oxide semiconductors, metal NPs (solid inorganic matters), or molecular linkers, organics, proteins (soft matters) [25].

Gerengi et al., 2011 reported the corrosion inhibition by extract of Schinopsis lorentzii was intended for utilizing Tafel extrapolation, linear polarization, and EIS of low carbon steel with diverse concentrations of 1M HCl solution. The results showed that with the upsurge concentration of extract of Schinopsis lorentzii, efficacies of inhibition augmented, which performed as somewhat cathodic inhibitor. It was demonstrated that in the HCl atmosphere, the low carbon steel corrosion inhibition resulted by Schinopsis lorentzii extract.

In view of the inhibition efficacy, the electrochemical procedures i.e., LPR, TP, and EIS utilized in this study display comparable outcomes on Schinopsis lorentzii extract. The extract of Schinopsis lorentzii might be utilized as a substitute in acid pickling and acidization of mild steel for toxic chemical inhibitors, and it is also an ecologically and naturally occurring material [26].

Based on three corrosion inhibitors: mercapto-benzimidazole, mercapto-benzothiazole, and benzotriazole loaded with halloysite nanotubes, Joshi et al., 2013 developed long-lasting anticorrosive coatings for steel. The oil-based alkyd paint at 5−10 wt%, together with corrosion healing and continued agent release in the coating defects, were admixed with the inhibitors' loaded tubes. In the anticorrosion efficacy of the coatings, the sluggish inhibitor proclamation (20−30 h) at flaw facts produced an outstanding upgrade. At nanotube ends the development of release stoppers with copper-inhibitor complexation and urea-formaldehyde copolymer, the time growth has been attained of anticorrosion agent release. Through the micro scanning of growth of corrosion current through reviewing paint adhesion and microscopy examination, in a 0.5 M NaCl solution, the corrosion protection efficacy on ASTM A366 steel plates was tested. The finest shield was attained via benzotriazole and halloysite/Mercapto-benzimidazole inhibitors. Consequently, by the longer release of inhibitors, and added an upsurge in corrosion efficacy provided through the development of stoppers by a copolymer of urea−formaldehyde.

For the growth of additional useful coatings, for example, antimicrobial, antifouling, and flame retardant that helpful from the controlled release and encapsulation of the vigorous agents, this method might be useful [27].

Amongst numerous choices for industrial-scale CO_2 capture, RTIL systems are likewise existence verified as a probable replacement to report the disadvantages of aqueous alkanolamine grounded advanced method. To hinder the existence of corrosion, these novel arrangements appear as an improved substitute. In water-based chemical absorption methods, oversight of the aqueous phase marks the elimination of plausible oxidizing classes primarily accountable for corrosion. By the steadily searching the outcome of process temperature, amine/RTIL type, in flue gas absence/presence of O_2, loading of CO_2, besides the effect of amount of water, the corrosion procedure in RTIL mixtures comprised hydrophilic RTIL ([EMIM][Otf], [BMIM][BF$_4$], and [EMIM][BF$_4$]) and alkanolamines have been examined. The adsorption of toxic materials on corrosion inhibitors was done by various analytical tools such as EDX, SEM, and LPR.

Materials Research Forum LLC
https://doi.org/10.21741/9781644901496-2

An extremely operative and cost-effective scheme is vital, which may hamper the CO_2 flow hooked on the atmosphere to handle anthropogenic GGE. The outcomes were associated through aqueous alkanolamine systems to test their suitability and to study the corrosion rate in alkanolamine−RTIL solutions of carbon steel 1020.

The experimental data showed the following conclusions:

- [BMIM][BF$_4$] and [EMIM][BF$_4$] ionic liquids established a comparable trend in grouping with alkanolamines and displayed outstanding control of corrosion in the pure state. However, because of the attendance of acidic impurities, clean [EMIM][Otf] was moderately corrosive in the electrode's direction surface, mainly.

- An advanced gas loading additionally enhanced the corrosion inhibition competence of amine− RTIL blends in contradiction of aqueous amine systems.

- The damaging consequence of the liquid fillings shields the metal surface by the development of a protecting film, which was demonstrated by EDX analysis. Furthermore, on the protecting ability of amine−RTIL mixtures contrary to corrosion, the absence/presence of H_2O and O_2 did not enclosure harmful effects at all [28].

In corrosive media like ionic liquid, organic, and aqueous media, Chatterjee et al., 2014 demonstrated the nanostructured silicon electrochemical steadiness permitted thru heterogeneous boundaries of conformal few-layered graphene. The passivation (d = 0.35 nm) of few-layered graphene preserved through direct gas-phase, which is transportable to P-Si materials besides Si-NPs. For Si-NPs, the unpassivated Si-NPs impulsively liquify, whereas the graphene-passivated Si corresponds to endure physical deterioration. They are empowering these materials for electrochemical supercapacitors for P-Si, the electrochemical steadiness thru extensively diverse electrolytes together with NaOH. They emphasized the straight establishment of nanostructured silicon on few-layered graphene transversely energy storage, electronics, energy conversion, biological systems, and detection because the superficial interactions only command the use of nanoscale silicon in varied submissions resultant to formulate heterogeneous on-chip boundaries which could preserve steadiness in even the utmost sensitivity towards surroundings.

For usage in a variety of chemical and biological surroundings, this work highlighted on atomically thin graphene passivation layers to permit Si-NPs platform for silicon material that is chemically versatile, and its manufacturing is cost effective. Particularly, to assimilate energy stowage hooked on present technical substantial contexts, the view of on-chip increased power energy stowage permits a comprehensive podium, applicable for detection, solar conversion, and electronics, besides emergent podiums, for example, implantations and recyclable systems [29].

Baek et al., 2012 reported the synthesis of green corrosion inhibitors using three-step reactions, i.e. click coupling reaction, esterification reaction, and nucleophilic substitution reaction using n-dodecyl succinate spinoffs comprising two DSE or HSE. The formation of succinate derivatives has been characterized using many analytical tools such as FT-IR, ^{1}H NMR, and EDX analysis. The weight loss and ASTM D665 methods were utilized to investigate the corrosion averting possessions of the ending harvests. It was found that because of the occurrence of succinate derivatives in paraffin oil, the strength of corrosion protection possessions was not principally determined by the functional groups or their length, but it was determined via the attentiveness of alkyl chains.

In the synthesis mechanism, the succeeding reactions occur:

- The first reaction occurs from sodium azide to an alkyl halide is the nucleophilic substitution reaction.

- Second, amongst propargyl alcohol and alkyl azide is the click coupling reaction.

- Third, amongst triazole alcohol and dodecyl succinic anhydride is the esterification reaction.

- The results showed that the DSE showed slighter low corrosion inhibition efficiency at low concentrations than the HSE, which shows that on the steel surface, the -COOH groups can improve the development of the lubricant layer to some amount. Alternatively, the corrosion inhibition efficacy augmented on a growing quantity of DSE and HSE spinoffs. The DSE and HSE spinoffs displayed the same efficacies towards inhibition of corrosion at high concentrations. The outcomes show the efficiency of HSE and DSE towards green corrosion inhibitors [30].

Hasib-ur-Rahman et al., 2013 reported aqueous alkanolamine-based CO_2 capture procedures as a major concern for corrosion. In aqueous monoethanolamine solvents, for the control of the corrosion process, they examined the competence of an almost nonvolatile and thermodynamically steady RTILs. The [emim][DCA], [emim][Otf], [emim][tosylate], and [emim][acetate] were the four imidazolium-based RTILs having ethyl side chains which have been selected for the purpose of corrosion. In industrial installations, for the process of corrosion, carbon steel 1020 which has been utilized as a test material is extensively utilized as a construction material. Electrochemical corrosion tests for measuring corrosion current were carried out via the LPR method, therefore, permitting the ensuing rate calculation of corrosion by the Tafel fit technique. In alkanolamine-based advanced CO_2 capture schemes, the [emim][acetate] is the utmost accomplished beyond the verified ionic liquids of restoring the various working difficulty of corrosion.

Sustainable Corrosion Inhibitors   Materials Research Forum LLC
Materials Research Foundations**107** (2021) 30-45   https://doi.org/10.21741/9781644901496-2

Through the oxidants hindering drive in the direction of the metal surface, these RTILs look to check the corrosion method if possible, owing to their higher viscosity. Though, thru in view of the ionic liquids that have an improved attraction for carbon steel and added methodical analysis is essential [31].

Pathan et al., 2013 reported the surface covering manufacturing is worldwide intricate in emerging biodegradable, valuable, and VOCs free covering schemes, which exhibit auspicious corrosion inhibition and physicomechanical recital. In this, the author reported the fabrication of BMF modified SA-BMF. The molecular weight and mechanical illumination of SA and SA-BMF resin were examined thru GPC methods and ^{1}H NMR, ^{13}C NMR, and FT-IR spectroscopy. The physico-mechanical and physico-chemical possessions were verified by standard procedures. The TGA analysis was utilized to quantity initial temperature decay. The glass transition temperature and curing study were measured using the thru DSC. The contact angle was used to determine the hydrophobicity of the coatings. The EIS and polarization methods were utilized to examine anticorrosion recital of SA-BMF. Inclusively, the waterborne SABMF covering showed bend tests (flexibility retentive 1/8 in.), scratch resistance, and developed impact resistance (150 lb/in.), together with decent anticorrosive recital and antibacterial action.

For the surface covering manufacturing, the PDP contact angle and EIS studies exposed virtuous rust prevention recital in contrast to alkaline, acid, and tap water media make them appropriate [32].

For antimicrobial and anticorrosion action, Oguzie et al., 2013 reported capsicum frutescens (CF) extracts. In acidic media, the antimicrobial efficiency of ethanol, methanol, petroleum spirit, and water extract correspondingly contrary to the corrosion-associated SRB, Desulfotomaculum classes, was measured utilizing the agar disc diffusion method, whereas the anticorrosion significance on low carbon steel of the ethanol extract was deliberated practically utilizing impedance, polarization, and gravimetric methods. Because of the accomplishment of the phytochemical ingredients existing within the structure, together with 39.2% saponins, 0.4% tannins, and 8.8% alkaloids, CF extract efficiently prevented SRB growth and corrosion equally. The antimicrobial outcome consequences from vital metabolic purposes and disturbance of the growth of the SRB, whereas the corrosion course was prevented thru adsorption on the steel surface of organic extract material. To the corrosion preventing the act of the extract, dihydro-capsaicin, and capsaicin, and afforded molecular level visions on their separate contributions, MD imitations were completed to demonstrate theoretically, the adsorption performance and electronic assembly of the lively alkaloidal ingredients extract of CF [33].

For the generation of VOCs free coatings and paints, Rahman et al., 2017 reported HBA−BMF− Fe_3O_4 spherical designed oleo alkyds which uphold decent fluidity, low viscosity and perform a vital part for the efficacy towards corrosion prevention. The pentaerythritol, SO, and phthalic anhydride were used to synthesize HBA (Hyperbranched alkyd). For the synthesis of the HBA−BMF− Fe_3O_4 nanocomposite for anticorrosive coatings, a sonication method was used to disperse the Fe_3O_4 NPs (magnetite) in HBA−BMF. To assess morphological, structural, thermal, physico-mechanical, anticorrosive, and electrochemical possessions of these coatings, the ASTM techniques were utilized. The coverings of HBA−BMF and HBA−BMF−Fe_3O_4 nanocomposite displayed virtuous physicomechanical possessions and suppleness. The outcomes of electrochemical corrosion studies demonstrate that in comparison to HBA−BMF, the HBA−BMF−Fe_3O_4 nanocomposite coverings display greater corrosion prevention recital having 1.0×10^{-4} mpy of at 107 Ω of impedance value [34].

Upadhyayula et al., 2017 reported graphene-modified PEI coatings as a key apprehension for bare steel in susceptibility to neighboring corrosion and structural submissions. HDG coatings suffer from the need for recurrent maintenance cycles, and difficulties associated with shorter service exist. To address these problems, Graphene-modified PEI coverings have been predictable as appropriate substitutes for corrosion inhibitors. Though, to comprehend the potential ecological effects of these coatings, general anxieties concerning the insinuations of nanomaterials make it necessary. To progress covering adsorption, the detection studies reveal that the probable influences insensitive to the usage of chemically or thermally functionalized graphene are tremendously sensitive to the conservation requirements and amenity lifespan of the covering [35].

A first-generation biofuel that presently exists as a biocomponent blended with types of gasoline reported by Matějovský et al., 2019. Combinations of gasoline and ethanol are elected as EGBs. To numerous nonmetallic and metallic resources due to moisture affinity and elevated polarity of ethanol, which significantly affect the possessions of the subsequent EGB, including their viciousness. Through corrosion inhibitors, the EGBs corrosion aggressiveness might be diminished. The E10, E25, E60, and E85 fuels are the three corrosion inhibitors for the inhibition of corrosion which have been reported on mild steel. The verified inhibitors were DET, A and two compound inhibitors comprising DBS, PA, and ODA. For the DETA concentration of 100 mg·L^{-1}, the inhibitory efficacy was found to be 98% in the E60 fuel [36].

For the high-density petroleum resulting fuels, Liang et al., 2015 reported renewable substitutes as energy-dense terpene biofuels. Coastal sediment/seawater developments incubated underneath simple anaerobic circumstances and were revised with TDF or a mixture of consequential petroleum hydrocarbons [37].

AlHarooni et al., 2016 emphases on evaluating 6 investigative methods to examine the deprivation state of many MEG solutions comprising FFCI and MDEA that were thermally treated to 135 °C - 200 °C for the corrosion and gas hydrate establishment inside gas pipelines which are two key problems towards flow assurance. The investigative methods appraised are electrical conductivity, pH measurement, ion chromatography (IC), alteration in physical features, HPLC-MS, and GHIP (using CH_4 gas at 50 to 300 bar pressure with MEG solutions having 20 wt %). This study mainly focusses on the effect of degraded MEG on corrosion inhibition and gas hydrate because of the deficiency of the value controller [38].

Matějovský et al., 2018 described the bioethanol might be partake a significant stimulus proceeding towards complete thermo-oxidative steadiness and further hooked on types of n gasoline, expressively varies the chemical and physical possessions of the ensuing fuels. The authors synthesized ethanol–gasoline blends (EGBs) E10, E25, E40, E60, and E85 contingent on their initiation period and were artificially oxidized [39].

Similarly, inside a failed nuclear waste container, to inhibit corrosion of consumed nuclear fuel, Liu et al., 2016 reported a combination of H_2 from H_2O_2 disintegration and many reactions concerning liquified H_2 formed whichever thru α radiolysis otherwise thru the deterioration of the steel container vessel [40].

Conclusion

In this chapter, various materials have been discussed which were utilized as corrosion inhibitors to prevent metal deterioration. The inhibitors showed high adsorption efficiencies towards toxic substituents adhere to on the metal surface and cause corrosion. These materials are highly cost effective, elevated potential, clean, renewable, sustainable source of energy, and cost-effective owing to the population growth and fast progress in response to the ecological concerns and ever-increasing global energy request. Therefore, they have a future perspective the energy conservation and deterioration prevention and fascinate the researcher's attention.

Abbreviations

VOCs = Volatile organic compounds

GO = Graphene oxide

RGO = Reduced graphene oxide

EIS = Electrochemical impedance spectroscopy

RTIL = Alkanolamine/room-temperature ionic liquid

LPR = Linear polarization resistance

SEM = Scanning electron microscopy

EDX = Energy-dispersive X-ray spectroscopy

GGE = Greenhouse gas emissions

P-Si = Electrochemically etched porous silicon

Si-NPs = Silicon nanoparticles

DSE = triazole groups

DSE = two triazole groups

HSE = one triazole groups

LPR = Linear polarization resistance

BMF modified SA-BMF = Butylated melami-e formaldehyde -modified soy alkyd

GPC = Gel permeation chromatography

TGA = Thermogravimetric analysis

DSC = Differential scanning calorimetry

EIS = Electrochemical impedance spectroscopy

SRB = Sulfa-e reducing bacteria

MD = Molecular dynamics

Mpy = Mils per year

PEI = Poly(ether imide)

EGBs = Ethanol−gasoline blends

DETA = Diethylene triamine

DBS = Dibenzyl sulfoxide

PA = Propargyl alcohol

ODA = Octadecyl amine

HPLC-MS = High performance liquid chromatography−mass spectroscopy

GHIP = Gas hydrate inhibition performance

Acknowledgements

The authors gratefully acknowledge the support from the Ministry of Human Resource Development Department of Higher Education, Government of India under the scheme of Establishment of Centre of Excellence for Training and Research in Frontier Areas of Science and Technology (FAST), for providing the necessary financial support to perform this study vide letter No, F. No. 5–5/201 4–TS.Vll. Dinesh Kumar is also thankful DST, New Delhi for financial support to this work (sanctioned vide project Sanction Order F. No. DST/TM/WTI/WIC/2K17/124(C).

References

[1] O.U. Rahman, M. Kashif, S. Ahmad, Nanoferrite dispersed waterborne epoxy-acrylate: Anticorrosive nanocomposite coatings, Prog. Org. Coat. 80 (2015) 77–86. https://doi.org/10.1016/j.porgcoat.2014.11.023

[2] O.U. Rahman, S.I. Bhat, H. Yu, S. Ahmad, Hyperbranched soya alkyd nanocomposite: A sustainable feedstock-based anticorrosive nanocomposite coatings, ACS Sustain. Chem. Eng. 5 (2017) 9725–9734. https://doi.org/10.1021/acssuschemeng.7b01513

[3] D. Borisova, H. Möhwald, D.G. Shchukin, Influence of embedded nanocontainers on the efficiency of active anticorrosive coatings for aluminum alloys part II: Influence of nanocontainer position, ACS Appl. Mater. Interfaces, 5 (2013) 80–87. https://doi.org/10.1021/am302141y

[4] T. Balakrishnan, S. Sathiyanarayanan, S. Mayavan, Advanced anticorrosion coating materials derived from sunflower oil with bifunctional properties, ACS Appl. Mater. Interfaces, 7 (2015) 19781– 19788. https://doi.org/10.1021/acsami.5b05789

[5] S.I. Bhat, Y. Ahmadi, S. Ahmad, Recent advances in structural modifications of hyperbranched polymers and their applications, Ind. Eng. Chem. Res. 57 (2018) 10754–10785. https://doi.org/10.1021/acs.iecr.8b01969

[6] A. Ghosal, O.U. Rahman, S. Ahmad, High-performance soya polyurethane networked silica hybrid nanocomposite coatings, Ind. Eng. Chem. Res. 54 (2015) 12770–12787. https://doi.org/10.1021/acs.iecr.5b02098

[7] M. Boruah, P. Gogoi, B. Adhikari, S.K. Dolui, Preparation and characterization of Jatropha Curcasoil based alkyd resin suitable for surface coating, Prog. Org. Coat. 74 (2012) 596–602. https://doi.org/10.1016/j.porgcoat.2012.02.007

[8] S. Pathan, S. Ahmad, Synthesis, characterization and the effect of the s-triazine ring on physico-mechanical and electrochemical corrosion resistance performance of waterborne castor oil alkyd, J. Mater. Chem. A. 1 (2013) 14227–14238. https://doi.org/10.1039/c3ta13126b

[9] M. Irfan, S.I. Bhat, S. Ahmad, Reduced graphene oxide reinforced waterborne soy alkyd nanocomposites: Formulation, characterization, and corrosion inhibition analysis, ACS Sustain. Chem. Eng. 6 (2018) 14820-14830. https://doi.org/10.1021/acssuschemeng.8b03349

[10] M. Bouklah, B. Hammouti, T. Benhadda, M. Benkadour, Thiophene derivatives as effective inhibitors for the corrosion of steel in 0.5 M H_2SO_4, J. Appl. Electrochem. 35 (2005) 1095–1101. https://doi.org/10.1007/s10800-005-9004-z

[11] A.S. Fouda, A.A. Al-Sarawy, E.E. El-Katori, Pyrazolone derivatives as corrosion inhibitors for C-steel HCl solution, Desalination, 201 (2006) 1–13. https://doi.org/10.1016/j.desal.2006.03.519

[12] A. Fiala, A. Chibani, A. Darchen, A. Boulkamh, K. Djebbar, Investigations of the inhibition of copper corrosion in nitric acid solutions by ketene dithioacetal derivatives, Appl. Surface Sci. 253 (2007) 9347–9356. https://doi.org/10.1016/j.apsusc.2007.05.066

[13] **V.R. Evans, The corrosion and oxidation of metals, Second Supplementary Volume, National Library of Australia, 1976.**

[14] B.E. Rani, B.B.J. Basu, Green inhibitors for corrosion protection of metals and alloys: An overview, Int. J. Corros. 2012 (2012). https://doi.org/10.1155/2012/380217

[15] A.Y. El-Etre, Natural honey as corrosion inhibitor for metals and alloys. I. Copper in neutral aqueous solution, Corros. Sci. 40 (1998) 1845–1850. https://doi.org/10.1016/S0010-938X(98)00082-1

[16] A.Y. El-Etre, Inhibition of aluminum corrosion using Opuntia extract, Corros. Sci. 45 (2003) 2485– 2495. https://doi.org/10.1016/S0010-938X(03)00066-0

[17] M. Kliskic, J. Radoservic, S. Gudic, V. Katalinic, Aqueous extract of Rosmarinus officinalis L. as inhibitor of Al- Mg alloy corrosion in chloride solution, J. Appl. Electrochem. 30 (2000) 823–830. https://doi.org/10.1023/A:1004041530105

[18] O.K. Abiola, N.C. Oforka, E.E. Ebenso, N.M. Nwinuka, Eco-friendly corrosion inhibitors: The inhibitive action of Delonix Regia extract for the corrosion of aluminium in acidic media, Anti-Corros. Method M. 54 (2007) 219–224. https://doi.org/10.1108/00035590710762357

[19] A.Y. El-Etre, Khillah extract as inhibitor for acid corrosion of SX 316 steel, Appl. Surf. Sci. 252 (2006) 8521–8525. https://doi.org/10.1016/j.apsusc.2005.11.066

[20] E.E. Ebenso, U.J. Ekpe, Kinetic study of corrosion and corrosion inhibition of mild steel in H_2SO_4 using Carica papaya leaves extract, West Afr. J. Biol. Appl. Chem. 41 (1996) 21–27.

[21] E.E. Ebenso, U.J. Ibok, U.J. Ekpe, Corrosion inhibition studies of some plant extracts on aluminium in acidic medium, Transactions of the SAEST, 39 (2004) 117–123.

[22] U.J. Ekpe, E.E. Ebenso, U.J. Ibok, Inhibitory action of Azadirachta indica leaves extract on the corrosion of mild steel in H_2SO_4, West Afr. J. Biol. Appl. Chem. 37 (1994) 13–30.

Sustainable Corrosion Inhibitors
Materials Research Foundations **107** (2021) 30-45
Materials Research Forum LLC
https://doi.org/10.21741/9781644901496-2

[23] F. Zucchi, I.H. Omar, Plant extracts as corrosion inhibitors of mild steel in HCl solutions, Surface Technol. 24 (1985) 391–399. https://doi.org/10.1016/0376-4583(85)90057-3

[24] S.A. Umoren, O. Ogbobe, I.O. Igwe, E.E. Ebenso, Inhibition of mild steel corrosion in acidic medium using synthetic and naturally occurring polymers and synergistic halide additives, Corros. Sci. 50 (2008) 1998– 2006. https://doi.org/10.1016/j.corsci.2008.04.015

[25] X. Gong, G. Liu, Y. Li, D.Y.W. Yu, W.Y. Teoh, Functionalized-graphene composites: Fabrication and applications in sustainable energy and environment, Chem. Mater. 28 (2016) 8082-8118. https://doi.org/10.1021/acs.chemmater.6b01447

[26] H. Gerengi, H.I. Sahin, Schinopsis lorentzii extract as a green corrosion inhibitor for low carbon steel in 1 M HCl solution, Ind. Eng. Chem. Res. 51 (2011) 780-787. https://doi.org/10.1021/ie201776q

[27] A. Joshi, E. Abdullayev, A. Vasiliev, O. Volkova, Y. Lvov, Interfacial modification of clay nanotubes for the sustained release of corrosion inhibitors, Langmuir, 29 (2012) 7439-7448. https://doi.org/10.1021/la3044973

[28] M. Hasib-ur-Rahman, H. Bouteldja, P. Fongarland, M. Siaj, F. Larachi, Corrosion behavior of carbon steel in alkanolamine/room-temperature ionic liquid based CO_2 capture systems, Ind. Eng. Chem. Res. 51 (2012) 8711-8718. https://doi.org/10.1021/ie2019849

[29] S. Chatterjee, R. Carter, L. Oakes, W.R. Erwin, R. Bardhan, C.L. Pint, Electrochemical and corrosion stability of nanostructured silicon by graphene coatings: Toward high power porous silicon supercapacitors, J. Phys. Chem. C, 118 (2014) 10893-10902. https://doi.org/10.1021/jp502079f

[30] S.Y. Baek, Y.W. Kim, S.H. Yoo, K. Chung, N.K. Kim, J.S. Kim, Synthesis and rust preventing properties of dodecyl succinate derivatives containing triazole groups, Ind. Eng. Chem. Res. 51 (2012) 9669-9678. https://doi.org/10.1021/ie300316f

[31] M. Hasib-ur-Rahman, F. Larachi, Prospects of using room-temperature ionic liquids as corrosion inhibitors in aqueous ethanolamine-based CO_2 capture solvents, Ind. Eng. Chem. Res. 52 (2013) 17682-17685. https://doi.org/10.1021/ie401816w

[32] S. Pathan, S. Ahmad, s-Triazine ring-modified waterborne alkyd: Synthesis, characterization, antibacterial and electrochemical corrosion studies, ACS Sustain. Chem. Eng. 1 (2013) 1246-1257. https://doi.org/10.1021/sc4001077

[33] E.E. Oguzie, K.L. Oguzie, C.O. Akalezi, I.O. Udeze, J.N. Ogbulie, V.O. Njoku, Natural products for materials protection: Corrosion and microbial growth

inhibition using Capsicum frutescens biomass extracts, ACS Sustain. Chem. Eng. 1 (2012) 214-225. https://doi.org/10.1021/sc300145k

[34] O.U. Rahman, S.I. Bhat, H. Yu, S. Ahmad, Hyperbranched soya alkyd nanocomposite: A sustainable feedstock-based anticorrosive nanocomposite coatings, ACS Sustain. Chem. Eng. 5 (2017) 9725-9734. https://doi.org/10.1021/acssuschemeng.7b01513

[35] V.K. Upadhyayula, D.E. Meyer, V. Gadhamshetty, N. Koratkar, Screening-level life cycle assessment of graphene-poly (ether imide) coatings protecting unalloyed steel from severe atmospheric corrosion, ACS Sustain. Chem. Eng. 5 (2017) 2656-2667. https://doi.org/10.1021/acssuschemeng.6b03005

[36] L. Matějovský, J. Macák, O. Pleyer, P. Straka, M. Staš, Efficiency of steel corrosion inhibitors in an environment of ethanol–gasoline blends, ACS Omega, 4 (2019) 8650-8660. https://doi.org/10.1021/acsomega.8b03686

[37] R. Liang, B.G. Harvey, R.L. Quintana, J.M. Suflita, Assessing the biological stability of a terpene-based advanced biofuel and its relationship to the corrosion of carbon steel, Energy & Fuels, 29 (2015) 5164-5170. https://doi.org/10.1021/acs.energyfuels.5b01094

[38] K. AlHarooni, D. Pack, S. Iglauer, R. Gubner, V. Ghodkay, A. Barifcani, Analytical techniques for analyzing thermally degraded monoethylene glycol with methyl diethanolamine and film formation corrosion inhibitor, Energ. Fuels 30 (2016) 10937-10949. https://doi.org/10.1021/acs.energyfuels.6b02116

[39] L. Matějovský, J. Macák, M. Pospíšil, M. Staš, P. Baroš, A. Krausová, Study of corrosion effects of oxidized ethanol–gasoline blends on metallic materials, Energ. Fuels 32 (2018) 5145-5156. https://doi.org/10.1021/acs.energyfuels.7b04034

[40] N. Liu, L. Wu, Z. Qin, D.W. Shoesmith, Roles of radiolytic and externally generated H_2 in the corrosion of fractured spent nuclear fuel, Environ. Sci. Technol. 50 (2016) 12348-12355. https://doi.org/10.1021/acs.est.6b04167

Sustainable Corrosion Inhibitors   Materials Research Forum LLC
Materials Research Foundations 107 (2021) 46-69   https://doi.org/10.21741/9781644901496-3

Chapter 3

Green Corrosion Inhibitors from Biomass and Natural Sources

A.N. Grassino[1]*, I. Cindrić[2], J. Halambek[2]

[1]Faculty of Food Technology and Biotechnology, University of Zagreb, Pierottijeva 6, 10000 Zagreb, Croatia

[2]Karlovac University of Applied Sciences, Trg J. J. Strossmayera 9, 47000, Karlovac, Croatia

* aninc@pbf.hr

Abstract

Considering the fact that corrosion of metals and alloys presents a significant problem all over the world, the one of most recently utilized approaches to combat this problem necessitates the researches for employment of new materials, which satisfied the green chemistry idea. In this connection, the development of sustainable corrosion inhibitors is highly demanded due to the increasing of awareness of green chemistry principles not only in corrosion discipline, but also in all branches of science and technology. Due to natural and biological origin as well as their eco-friendly extraction, the plant materials and biomass derived from various waste sources could be applied as beneficial substances for metals and alloys protection in different corrosion environment. Therefore, this work reports the main findings regarding their employments as green anticorrosion substances.

Keywords

Natural Sources, Biomass Waste, Green Corrosion Inhibitors, Metal Surface, Corrosion Media

Contents

1.   Introduction

The corrosion of metals and alloys is a serious industrial issue all over the world, which could be minimized or stopped by suitable approaches. One of them involves the application of corrosion inhibitors [1], i.e. the substances which are able to slow down the corrosion reactions when added in low concentration. Due to the fact that corrosion involves the movement of the metal ions into the aggressive solution from the more (anode) to less (cathode) active areas, with an ionic current in the solution, and an electronic current in the metal, the addition of inhibitors reduced the corrosion rate by increasing or decreasing the anodic and/or cathodic reactions [2]. These molecules generally contained heteroatoms, such as oxygen, nitrogen and sulfur, through are adhered onto the metal, forming a barrier for further corrosion attack. Although, the literature reports [1,3] showed that various organic compounds are successful anticorrosion substances for metals and alloys in different environments, unfortunately their use is limited due to their toxic nature and expensive synthesis.

Nowadays, with increasing of "green chemistry" concept in all branches of science, technology and engineering, a set of principles that causes reduction of environmental pollution, and increase of eco-friendly chemicals are influenced also in the corrosion science, regarding the employment of inhibitors derived from natural resources and biomasses [4]. In that connection, the plants as natural products have been recently proposed [5,6] as a green corrosion inhibitor due to its cheap, environmentally acceptable, abundant, readily available, and effective anticorrosion properties. There is a rich source of bioactive compounds such as tannins, phenols, flavonoids, acids, alkaloids, catechins, terpenoids, amino acid and proteins, polysaccharides and vitamins [4,5].

The numerous plant extracts have been prepared from various parts of plants, including leaf, root, stem, bark, pulp, fruit, etc., and investigated as green corrosion inhibitors for many metallic surface [7,8]. Among various phytochemicals in plant extracts seem that polyphenols classes satisfy the most of the demands of anticorrosion inhibitors [9], including that they are less toxic than other natural compounds, for instance alkaloids. Due to the fact that phytochemicals are extracted from various parts of plants, the proportion of constituents responsible for anticorrosion performance differ among them, and consequently attribute to different inhibition efficiency. It is certain that plant extracts and oils gained from leaf emerge to be an effective type of inhibitors to corrosion and could be successfully used on an industrial level, due to the high inhibition efficiency, up to 88 % [10]. However, the plant extracts could not resist high temperatures, and are not efficient in extremely aggressive environments [7].

Although efficient in anticorrosion activity, their exact mechanism is difficult to explain, due to its complex chemical structure. It could be anodic, cathodic or mixed. Usually, the anodic acted by creating a defensive oxide layer on the superficial of the metal, while cathodic ones are capable to reduce the cathodic reaction due to its selective precipitation on the cathodic areas of metal. When the inhibitors are able to provide both, cathodic and anodic actions, which is the one of usual characteristic of the numerous of green corrosion inhibitor then they could be classified as mixed types. Their adsorption on metallic surface is frequently associated by the presence of the double or triple bonds or V and VI groups of elements with free electron couples, i.e. N, P, S and O. It is also important to mention that certain types of inhibitors are strongly specific for one type of metal or alloy, and are not able to provide efficient inhibition of others.

Due to the fact that phytochemicals in plant extracts are the mixture of various constituents, it is necessary to provide the efficient qualitative or quantitative examination of plant extracts, in order to gain the idea which of the compound(s) is responsible for anticorrosion activity on metals or its alloys. Although, some attempts of corrosion scientists [10-14] have been made to connect the anticorrosion activity with certain chemical characterization of plant extracts, further investigations should be done. For instance, did the individual components act synergistically or antagonistically? In addition, if the active component showed by itself a very high degree of inhibition, particularly in aggressive medium, should it be isolated from the mixture? If the plant extracts could not withstand very high temperature (decomposition of come components), do scientists need to find and isolate a component that will show a high degree of inhibition even at higher applied temperatures? Thus, the future studies on application of plant extracts as corrosion inhibitors should be directed on isolation of the targeted, active molecule, which is responsible for the anticorrosive action.

Besides application of plant extracts as corrosion inhibitors, the employment of value-added compounds originated from un-utilized biomasses [15-32] as anticorrosion substances extended the investigations to another promising and superior candidates. Their chemical structures, similar to those of traditional inhibitors, and the fact that they are generated from low-cost and abundant waste biomasses opened new strategies towards to circular economy happen. Thus, in that connection the present paper summarized some of the natural resources and waste biomasses, which extracts are employed as green corrosion inhibitor.

2. Corrosion

The corrosion process [33] which could be chemical or electrochemical in nature requires at least two reactions that could occur in a corrosive environment. These reactions are classified as anodic and cathodic, and are defined by follow example, i.e. for metal M immersed in sulfuric acid solution the metal oxidation occurs through an anodic (1) and reduction through cathodic (2) reactions:

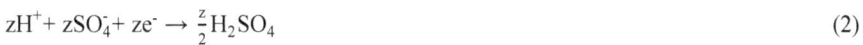

$$M \rightarrow M^{+z} + ze^- \qquad\qquad\qquad (1)$$

$$zH^+ + zSO_4^- + ze^- \rightarrow \tfrac{z}{2}H_2SO_4 \qquad\qquad\qquad (2)$$

Although, there are no unique classifications of the types of corrosion, the two common categories are established [34]: *i*) general (atmospheric, galvanic, high-temperature, liquid-metal, molten-salt, biological and stray-current, and *ii*) localized (crevice, filiform, pitting, oral, biological and selective leaching).

The tendency of a metal to corrode depends of various factors, which are mostly connected to metal itself, i.e. its composition and structure, degree of damage (macro or micro defects), and environmental conditions, i.e. type of corrosion medium, pH and acid content, the quantity of corrosion accelerators (oxygen, nitrate, sulfur, chloride and phosphate compounds in corrosion medium), as well as temperature and time of metal/metal alloys exposure to the corrosive medium, which significantly influenced on kinetics of metal oxidation and reduction of species in solution.

In order to avoid the corrosion damages, it is necessary to predict all possible failure, starting from metallic substrate to its final, desirable application, making technical, economic and environmental benefits. In that connection, the selection of appropriate metal or alloy, its testing, before and after interaction with environment, as well as analyses of environmental parameters involved in corrosion phenomena are important facts and must be considered.

Although, the different metallic materials needed to be protected from corrosion, the steel, aluminum, tin, zinc, copper and its alloy are frequently evaluated, due to their broad range of applications. For instance, the steels (carbon steel, alloy steel stainless steel and tool steel) have a wide range of applications (construction, manufacturing, transport, shipbuilding, energy and packaging) due to its highly durable, corrosion resistant, tensile and affordable properties. Tin as soft, silvery - white metal also possesses numerous properties, among which a un-toxicity and excellent resistance to aggressive attacks are more pronounce. Tin is usually integrated with other metals in order to make the alloys, which are possessed beneficial characteristic of tin. For instance, the tinplate (steel sheet coated with a layer of tin), efficiently expressed the corrosion resistance and low toxicity of tin with strength of steel. Although, the tin makes a minor proportion of tinplate, its addition to steel sheet made tinplate one of the frequently used materials for cans manufacture. Besides tinplate utilization in food and drinks industry (90 %), other applications involved cosmetics, fuel, oil, paints and other industry. Aluminium is used in the enormous number of products, such as cans, foils, kitchen utensils, window frames, beer kegs and airplane parts. It has a low density, great thermal conductivity and good corrosion resistance. It could be easy cast, manufactured and modeled. Aluminium is un-toxic and does not release any odors or taint products with which it is in contact, and thus in forms of aluminum foils and sheets is suitable in food packaging and pharmacy. The good corrosion resistance, excellent mechanical properties and capability to galvanize other metals made zinc currently one of the frequently utilized metals in various industrial branches. The copper as one of non-ferrous materials also possesses excellent properties, i.e. good electric conductivity, strength, ductility and corrosion resistance, which provided its broad application in electronics, energy, petrochemical, transportation, machinery, metallurgy and other industries.

2.1   Corrosion Inhibitors

Corrosion inhibitors are the substance that effectively reduces or prevents the corrosion of exposed metal in a corrosive environment [35]. Added in a small quantity, they are able to diminish the progression of corrosion, maintaining a protective, inhibitive film on metal superficial.

Corrosion inhibitors are classified as organic or inorganic compounds and over the last century, inorganic inhibitors such as chromates, vanadates, various nitrates etc. have been systematically used in various corrosive media [36] due to its remarkable inhibitory properties on different metals and alloys. However, their use is prohibited due to high toxicity, and thus the researches have been carried out on utilization of different organic compounds, especially newly synthesized heterocyclic molecules [3]. At the moment when it is understood that their preparation is extremely expensive, and when a large amount of hazardous organic solvents is used for their synthesis, the application of such compounds is avoided. Although, it has been proven that the vast majority of organic inhibitors protect the metal from the aggressive medium, mostly by adsorption, through atoms that possesses free electron pairs (N, S, O and P) or through double and triple bonds in the molecule, their utilization is also avoided due to the risks to human life and its environment [3].

The increasing of awareness of "green principles", i.e. ecological, economic and environmentally-friendly (e^3 concept) [37] in all fields of science and technology have been influenced also in the corrosion discipline. In that connection the use of traditional, synthetic organic compounds has been limited, and replace by green organic compounds, usually extracted by conventional or innovative processing technology, such as pulsed electric fields, ultrasound, microwave, supercritical fluid and high hydrostatic pressure [38]. The green corrosion inhibitors are renewable, bio-degradable and non-toxic, which made them a superior anticorrosion candidate then their synthetic counterparts. They are employment as inhibitors for various metallic materials in different media, such as HCl, H_2SO_4, H_3PO_4 and HNO_3. Nowadays, the plant extracts have acquired a numerous examination and application as green corrosion inhibitor, due to its natural origin. Additionally, the new compounds isolated from the under-utilized waste biomass as corrosion inhibitors have also gained the high attention due to transformation of waste into value-added product with anticorrosive activity, providing beneficial to the producer and environment.

In order to evaluate the inhibition activity of green corrosion inhibitors in various corrosive medium, the several standard analyses, including gravimetric test (weigh loss), and potentiodynamic polarization and impedance spectroscopy analyses are employed [4,5]. The weight loss test is the simplest technique used to calculate the corrosion rate, and it is bases on metal weight measurements before and after immersion in corrosive medium for a certain period. In comparison with weight loss analysis, the potentiodynamic polarization technique is able to provide a number of useful information regarding the rate of metal dissolution and its protection. For instance, current density, (i_{corr}), potential (E_{corr}) and Tafel slopes (β_a and/or β_c) are some of the polarization parameters frequently reported. Electrochemical impedance spectroscopy requires alternating current, and the output is Nyquist plot for charge transfer or diffusion control process. It could be used to determine polarization resistance R_p, which in turn is inversely proportional to the i_{corr} value. Besides these methods, the atomic force microscope (AFM) and scanning electron microscopy (SEM) are also two of powerful techniques applied to investigate the surface morphology, at nano to microscale. Both of them are capable to provide valuable information regarding the generation and progression of corrosion at the metal-solution interface. Nowadays, the computational studies and quantum chemical calculation are also in the progress, and they are so important for prediction of anticorrosion mechanism.

2.2   Corrosion inhibitors from plant sources

Plant extracts are extensively investigated since their main components have no notable side effect on human population and its surrounding, and since they are cheap, renewable and easy applicable. Besides these properties the extracts derived from plant sources contained the phytochemicals with similar characteristic and structure to organic corrosion inhibitors, i.e. they possess the heteroatom(s) and/or free electrons in the last orbital, making them as good candidates for inhibiting of metal corrosion in different media.

In order to employ plant extracts as anticorrosion substances for metals and alloys protection, the various methods and parameters for their preparation are available. For instance, the selection of the appropriate solvents is very important for efficient extraction of targeted compound(s). Although, water possesses the excellent properties (inexpensive, readily available, non-toxic, non-flammable and non-hazardous) in comparison with other solvents, the preparation of plant extracts requires mostly aqueous solution of organic solvents (ethanol and methanol), due to its better diffusion into plant cells. The second parameters, which must be considered is extraction temperature, and usually range from 60 to 80 °C. The extraction carried out from these optimal ranges could limit the solubility of targeted compounds or could cause its decomposition.

Nowadays, the uses of plant extracts in corrosion control are considered as green and environmental-friendly approach due to excellent solubility of plant phytochemicals in water. Organic acids, phenols and flavonoids, alkaloids, terpenoids amino acids and proteins, polysaccharides and vitamins are the usual constituents of a numerous of plant extracts [39]. Each of them could interact with a metal surface and retard the corrosion reactions due to the presents of polar functional groups such as -OH, -COOH, -COOC$_2$H$_5$, -OCH$_3$ etc. These groups are capable to increase the solubility of the compounds in a common corrosion media, such as water solution of inorganic acids, salts and alkali. The anticorrosion efficiency of mentioned constituents depends on the adsorption layers formed under particular conditions, and it is usually related with the origin and charge of the metal surface, the chemical composition of the inhibitor (functional groups, aromaticity, possible steric effects, etc.) and at least but not negligible the type of aggressive electrolyte [1]. As it is pointed above, the main mechanism of metals protection in aqueous media is the adsorption of inhibitor on metal surface. If the molecules of inhibitor existed in the system as protonated species (very acidic solutions), the inhibitor is capable to adsorbed on the metal surface by electrostatic feature, and the inhibitor act through so-called physisorption mechanism. [40]. If the inhibitor molecules existed in a solution as neutral molecules, they could donate their electron pairs to vacant orbital of a metal to make a covalent or co-ordinate types of bond, and in that case the chemisorption mechanism is achieved. Increase of system temperatures is favorable to this type of adsorption, i.e. the rises of temperature lead to increase of inhibitor activity. Although, the chemisorption mechanism of plant extracts has been observed on metal surface, it is the most likely that physisorption and chemisorption occurred on the same metal surface [41].

When the plant materials are considered as potential corrosion inhibitors, they are mostly evaluated on the basis of their bioactive constituents. They contain the particular functional groups, responsible for anticorrosion activity. For instance, the essential oil as concentrated hydrophobic liquids with pleasant odor and essence, usually extracted by steam distillation due to high volatility from various plant parts (flowers, stems, leaves, fruits, seeds and roots etc.) presented the complex mixture of constituents, at a quite different concentration. Their complex chemical compositions, among which aromatic and aliphatic compounds, and hydrocarbon terpenes (isoprenes) and terpenoids (isoprenoids) played a vital role in green anticorrosion performance. The essential oils also contained aldehydes (geranial, citronellal, etc.), ketone (pulegone, carvone, pinocarvone, etc.), alcohols (geraniol, citronellol, nerol, menthol, carveol, etc.), esters (linalyl acetate, isobornyl acetate, etc.) and ethers (1,8-cineole, menthofurane, etc.) [42], and as revealed Table 1 made them as suitable corrosion inhibitors for various metals and alloys, in different environments [43-59].

For instance, the corrosion inhibition efficiency of argan oil on corrosion of C38 steel in 1 M HCl solution obtained by potentiodynamic polarization analysis is 81 % with inhibitor concentration of 3 g/L [43]. Although the authors did not specify which of the components is responsible for the anticorrosion activity of oil, on the basis of calculated thermodynamic parameters they reported that the molecules from oil are physically adsorbed on the steel surface. The same technique is used for determination of inhibition activity of carob oil on steel (type C38) in 1 M hydrochloric acid [45]. Its inhibition action is conditioned by employed concentration, and with 0.5 g/L reached the 86.7 % efficiency. In spite of the fact that the chemical composition of carob oil is quite similar to argan oil (Table 1), the authors claim that molecules of inhibitors are adsorbed chemically on the steel surface. The fatty acids are also the main constituents of chamomile oil, which effectively blocked the steel corrosion in acidic medium depending on applied temperatures, i.e. 90 % at 298 K and 86 % at 328 K [47]. Limonene is the one of the components of several essential oils including verbena [44], fennel [48] and dill [54]. Although these investigations are performed in high acidic solutions, this cyclic monoterpene could not be protonated, and thus according to authors the molecules in the extracts are adsorbed by chemisorption on C38 steel, carbon steel and aluminum surfaces, respectively.

Sustainable Corrosion Inhibitors   Materials Research Forum LLC
Materials Research Foundations**107** (2021) 46-69   https://doi.org/10.21741/9781644901496-3

Table 1 Natural oils from plants as green corrosion inhibitors.

Natural oil	Main components	Material	Medium	Inhibitor concentration	Highest inhibition efficiency (%)	Refs.
Argan	Oleic acid, linoleic acid, spinasterol, γ-Tocopherol	C38 steel	1 M HCl	3 g/L	81.0	[43]
Verbena	Geranial, neral, limonene			5.0 g/L	85.5	[44]
Carob seed	Linoleic acid, palmitic acid, oleic acid			0.5 g/L	86.7	[45]
Rosemary	1,8-cineole	Steel	2 M H_3PO_4	10 g/L	73.0	[46]
Chamomile	Linoleic acid, oleic acid, palmitic acid		1 M HCl	2 g/L	90.0	[47]
Fennel	Limonene, β-pinene	Carbon steel	1 M HCl	3 mL/L	76.0	[48]
Garlic	Monoterpens			2.5 g/L	95.8	[49]
Menta	Menthol	Mild steel	1 M HCl	1 g/L	86.0	[50]
Pelargonium	Citronellol, geraniol, izometanon			4 ml/L	90.6	[51]
Laurel	1,8-cineole, methyl eugenol, linalool	Aluminium	1% CH_3COOH	45 mg/L	76.8	[52]
Basil	Linalool, eugenol, 1,8-cineol		0.5 M HCl	5.7 g/L	78.4	[53]
Dill	Carvone, limonene		1 M HCl	300 ppm	98.0	[54]
Cinnamon	Cinnamaldehyde, δ-cadinene	Copper	0.5 M H_2SO_4	150 ppm	89.6	[55]
Artemisia	Camphor, β-thujone	Tinplate	0.5 M HCl	5 g/L	81.0	[56]
Thymus	Borneol, α-terpineol			6 g/L	87.0	[57]
Lavander	Linalool, linalyl acetate	AA5754 alloy	3% NaCl	20 ppm	99.0	[58]
Laurel	1,8-cineol, methyl eugenol, linalool			50 ppm	82.4	[59]

Essential oils of different plants, especially those of the Mediterranean climate, such as laurel, lavender and basil, show very high protection of aluminium in acetic [52] and hydrochloric [53,54] acids, and neutral 3 % NaCl solution [58,59]. For instance, the degree of inhibition of essential oil extracted from basil oil [53] on aluminium in 0.5 M HCl solution gained by potentiodynamic polarization analysis is 90.1 % with oil concentration of 5.7 g/L. The authors reveled that linalool (2,6-dimethyl-2,7-octadien-6-ol) as major component of oil, and lower of eugenol and 1,8-cineole could be easily protonated, forming protective layer on aluminum by physisorption. Additionally, the lavander and laurel essential oils provided the excellent inhibition of aluminium alloy in 3 % NaCl against pitting corrosion due to presence of linalool [58,59]. The authors also assumed that besides physical adsorption on metal surface, the high inhibition efficiency (99.0% for lavender and 82.4% for laurel oils) is connected with the appearance of different adsorption centers, which are able to act synergistically to form a polymeric and coherent film on aluminium alloy. The study of rosemary oil on the corrosion of steel in 2 M H_3PO_4 solution [46] confirmed that its inhibition efficiency increases with increases of oil concentration, and the maximum value of 73 % is obtained with 10 g/L. Although 1,8-cineole as the major component of rosemary oil is responsible for anticorrosion activity, the authors suggested that intermolecular synergistic effect in which are participate all other active molecules in oil are also meritorious for excellent inhibition.

Besides essential oil, some of corrosion inhibitors derived from plant extracts [12,13,60-75] with the major components worthy for inhibition are shown in Table 2. For instance, the investigation of mild steel protection in 0.5 M sulphuric acid indicates that phytochemicals (methoxypyrazine, wine lactone and bisphenol A) found in the extracts of *Armoracia rusticana* [60] are responsible for its inhibition performance. The electrochemical measurements showed that this extract protected the mild steel with inhibition efficiency of 95.74 %, at concentration of 100 mg/L. The authors revealed that numbers of heteroatoms (N and O) in the molecules of methoxypyrazine, wine lactone and bisphenol A are able to create coordinate bonds, and thus through them are easily adsorbed on the surface of mild steel. Similarly, the *Azadirachta indica* extract contain a significant number of oxygen, sulfur and nitrogen atoms accountable for its inhibition activity [61]. Although, the authors reveled that it is difficult to assign which of the 300 compounds in *Azadirachta indica* extract is liable for the anticorrosion effect, and protection of mild steel. In the work of Chevalier et al. [12] is showed that anibine as major alkaloid (95.4 %) of *Aniba rosaeodora* extract is responsible for mild steel protection in high acidic solution by chemisorption. Although, the bamboo leaf extract contains flavonoids, amino acids and amylose, with O and N atoms in their functional groups (-OH, C-O, N-H and O-heterocyclic rings), the authors did not indicate which of the compounds is responsible for remarkable protection of steel in 1 M HCl (90.3 %) and 0.5 M H_2SO_4 (79.2 %) solutions after addition of 200 mg/L of extracts [62]. These authors [63] also studied the inhibition performance of *Ginkgo* leaves extract, rich of flavonoids, ginkgolides and amino acids, on steel surface in contact with HCl and H_2SO_4 solutions. They pointed out that the molecules from *Ginkgo* leaves extract as protonated in acid media formed the coordinate bond by partial transference of electrons from O and N atoms to vacant d orbits of Fe, providing excellent adsorption and protection of the steel surface. Similarly, the ginseng root extract as a mixture of glycoside, saponin, phenolic acids, alkaloids and lignin also acted as a good corrosion inhibitor of aluminium in HCl solution due to presence of heteroatoms in their structure [64]. According to the authors it is difficult to assign, which of the constituents is responsible for the inhibitive effect.

The black pepper extract also acted as excellent corrosion inhibitor of C38 steel in HCl medium, where the highest protection of 95.8 % are gained by 2 g/L [65]. At the same time, the authors have done the comparative study using piperine, which is isolated from black pepper powder. Its inhibition attains 99 % at 0.001 M concentration.

The influence of *Borage* flower extract on mild steel corrosion in 1 M HCl solution is studied by EIS and potentiodynamic polarization methods [66]. EIS results showed that maximum inhibition performance of 91 % is obtained at 800 ppm. In order to get the additional information regarding the anticorrosion mechanism of substances found in extracts, i.e. lactic acid, nicotinic acid and carotene, the authors are used the theoretical methods based on molecular simulations and quantum chemical calculations. They observed that all of three anticorrosion components, in neutral and protonated forms protected the steel surface with a flat molecular orientation. The other extract, such as that obtained from *Geissospermum* leave also showed a good protection of C38 steel in 1 M HCl, due to presence of alkaloid geissospermine. Its molecules are adsorbed onto the steel surface by physisorption of protonated species and chemisorptions of aromatic ring [67]. The authors also indicated that the high inhibition performance (over 90%) is gained by large size of geissospermine molecules, which covered the wide areas of steel surface and thus retarding the corrosion.

The adsorption and inhibitive efficiency of ethanolic extract of the leaves of *Chlomolaena odorata L.* on mild steel corrosion in acidic solution is probably attributed to the presence of different flavones [68]. However, the mutual effects of phytochemicals in the extract also may contribute to its high inhibition efficiency.

Aluminium corrosion in NaOH solution is studied after addition of aqueous extract of *Hibiscus sabdariffa* leaves [69]. The author revealed that the protection of aluminium surface is based on two phenomena, i.e. the first is the development of potentials at the metal-solution interface, and the second one is the presence of an oxide film. Considering the fact that the pH values of zero charge for aluminum oxide is 9.0-9.1, it is more likely that positively charged species in *Hibiscus* extract (thiamine and anthocyanins) tend to adsorb under the electrostatic attractions (physical adsorption mechanism). However, its constituents may also act synergistically to inhibit Al corrosion in the tested solutions. The *Saraca ashoka* extract [70], although contains the tannins, saponins and flavonoids, only epicatechin as flavonoids has ability to inhibit the corrosion of mild steel. As results showed the inhibition efficiency of 95.48 % is gained with 100 mg/L of inhibitor.

Overall, it is evident that most of the extracts contained the mixture of phytochemicals, and according to the authors it is difficult to interpret which of the substances possessed the anticorrosion activity. In this context, the future application of extracts derived from various plant materials as corrosion inhibitor must be focused on isolation of the specific active molecule, which is responsible for the anticorrosive properties.

Table 2. Plant extracts as green corrosion inhibitors

Plant extract	Main component	Material	Medium	Inhibitor concentration	Highest inhibition efficiency (%)	Type of adsorption	Refs.
Bamboo leaves	Flavonoids, amino acids	Steel	1 M HCl	200 mg/L	87.4	Mixed	[62]
			0.5 M H_2SO_4		78.8		
Ginko leaves	Flavonoids, ginkgolides, amino acids		1 M HCl	100 mg/L	91.4		[63]
			0.5 M H_2SO_4		81.6		
Geissospermum leaves	Geissospermine	C38 steel	1 M HCl	100 mg/L	90.0	Physisorption	[67]
Black peper	Piperine		1 M HCl	2 g/L	95.0	Chemisorption	[65]
Jasminum nudiflorum	Favonoids, phillyrin, verbascoside	Cold roled steel	1 M HCl	50 mg/L	93.0	Mixed	[71]
Ficus tikoua	Allantoin, 5 methoxypsoralen, methyl caffeate	Carbon steel	1 M HCl	200 mg/L	95.8	Chemisorption	[72]
Saraca ashoka	Epicatechin	Mild steel	0.5 M H_2SO_4	100 mg/L	95.5	Mixed	[70]

Materials Research Forum LLC

https://doi.org/10.21741/9781644901496-3

Armoracia rusticana	Methoxypyrazine, wine lactone, bisphenol A		$0.5\ M$ H_2SO_4	100 mg/L	95.4	Mixed	[60]
Chlomolaena odorata	Mixture of flavones		$0.5\ M$ H_2SO_4	5 % v/v	95.0 89.0	Physisorption	[68]
Artemisia annua	Artemisinin		$2\ M$ H_2SO_4	4 g/L	95.5	Physisorption	[13]
Azadirachta indica	Terpenoids-azadirachtin		$2\ M$ H_2SO_4	4 g/L	60.4 66.8 81.8	Physisorption	[61]
Aniba rosaeodora	Anibine		1 M HCl	200 mg/L	91.0	Chemisorption	[12]
Cuminum cyminum	Cuminaldehyde		1 M HCl	300 ppm	93.0	Physisorption	[73]
Borage flower	Lactic acid, nicotinic acid, carotene		1 M HCl	800 ppm	91.0	Mixed	[66]
Rosa canina	Marein, ascorbic acid, pectin, tannin		1 M HCl	800 ppm	86.0	Mixed	[74]
Fig leaves	Psoralene, vanillin, ß-sitosterol		2 M HCl	200 ppm	87.0	Physisorption	[75]
Ginseng	Unknown complex mixture	Aluminium	1 M HCl	50 % v/v	93.1 63.1	Physisorption	[64]
Hibiscus sabdariffa	Thiamine, niacine, ascorbic acid, anthocyanins		$0.5\ M$ NaOH	1.5 g/L	85.0	Physisorption	[69]

2.3 Corrosion inhibitors from biomass waste

Nowadays, the cost effectives prevention of metals and its alloys utilizing natural compounds is of a great priority for industries. Several authors have employed various valuable components derived from un-utilized biomasses to yield efficient protection of metallic surfaces. For instance, Gapsari et al. [15] applied the extracts derived from the honeycomb waste as the inhibitor of stainless steel, type 304, in 0.5 to 2.0 M H_2SO_4 solution. They found that corrosion rate of stainless steel increases up to 2000 mg/L of inhibitor, and decreases with a concentration of 3000 to 4000 mg/L. The highest inhibition efficiency of 97.29 % is reached by inhibitor concentration of 2000 mg/L in 0.5 M H_2SO_4. According to the authors, its inhibition performances is associated to the adsorption behavior of honeycomb extracts due to presence of quercetin, luteolin, vitexin, fisetin, isohamnetin, isoferulic, apigenin and pinobanksin.

The extracts from other waste materials, such as eggshell is found to significantly alter the corrosion resistance of austenitic stainless steel, type 316 [16]. The authors showed that the increase of inhibitor concentration hinders the formation of a passive film, and the highest protection of 94.74% is gained by weight loss analysis.

In the work of Montoya et al. [17] it is found that the two primers with organic corrosion inhibitors extracted from Pinus radiata bark, from the waste generated by the timber and paper industries could be used as inhibition of 1020 steel due to formation of tannin-Fe protecting coatings.

Cruz-Zabalegui et al. [18] have used a non-ionic gemini surfactant (N,N-diethylaminedialkyldiamide) synthesized from the fatty acids derived from avocado oil as CO_2 inhibitor of API X-52 steel. By the test performed at 50 °C, during 24 h with inhibitor concentration of 10 to 50 ppm, the authors pointed out that this compound assured good protection due to physical adsorption onto the steel. In the work of Liao et al. [19] is showed that the waste of lychee (peel and seed) is possible to recycle and used as environmentally friendly inhibitor of mild steel in 0.5 M HCl solution up to concentration of 600 mg/L to achieves the maximum protection.

Although, the *Prunus dulcis* (almond) is one of the frequently used fruits in human diet, its peels are thrown out as a waste material. However, the idea of utilizing this waste material for protection of mild steel in 0.1 M HCl solution is gained by Pal et al. [20]. They showed that methanolic and aqueous extracts of almond protected the mild steel with efficiency of 93 and 85 %. The more pronounced inhibition performance of methanolic than aqueous extract, the authors supported this by SEM and AFM analyses. By UV/Vis and FTIR spectroscopy analyses, the authors showed that methanol contain more phytochemicals than water, among which the isorhamnetin-3-O-rutinoside, chlorogenic acid and catechin are those responsible for mild steel inhibition. The authors also revealed that extracts are classified as mixed type, where the adsorption is neither pure physical nor pure chemical.

Tiwari et al. [21] used the ethyl acetate extract of *Musa Paradisica* peels for mild steel protection in hydrochloric and sulphuric acids. The results of electrochemical study strongly validate that extract of *Musa Paradisica* peels retards mild steel corrosion in both acidic medias, due to abundancy of heteroatoms (oxygen and nitrogen) and aromatic rings, which are adhered on the surface via chemical or electrostatic interaction with iron atoms. Although banana peel extract contains several bio-active compounds, each of them could contribute to some extent in the total inhibition performance of the extract. According to the authors the inhibition activity is primarily controlled by the gallocatechin molecule.

In the work of Soares Rodrigues [22], the biomass of the microalgae *Spirulina maxima*, a cyanobacterium that synthesizes high levels of protein is studied as a natural inhibitor of carbon steel in 1 M HCl. This biomass acted as a good corrosion inhibitor reaching an efficiency of 96.4 %, after 72 h of immersion with 100 mg/L of inhibitor. Furthermore, the chitosan extracted from exoskeletons of crustaceans in the sea food waste led to about 88 % of corrosion inhibition through chemical and physical types of adsorption onto carbon steel. The corrosion inhibition efficiency of extracellular polymeric substances (EPS) extracted from waste activated sludge on carbon steel in 3.64 % NaCl solution saturated with CO_2 at 25 °C is examined by Go et al. [24]. In comparison with waste activated sludge alone with EPS extracted from the waste showed that the better inhibition performance of around 80 % is gained by extracellular polymeric substances with its concentration of 1.000 mg/L. This compound acted as a mixed-type of inhibitor with O-H, N-H, C-N, C=O and C-H groups, which contributed to adsorption on steel.

In the work of Dehghani et al. [25], the Chinese gooseberry fruit shell is used as green corrosion inhibitors of carbon steel, in HCl solution. Its aqueous extract composed of sucrose, maltose and folate, with concentration of 1000 ppm assured 92 % of inhibition, after 2.5 h of steel immersion. The results obtained by EIS are also supported by molecular simulation findings, which point out that among three major organic constituents the folate possessed the higher inhibition activity than sucrose and maltose. As authors declared the hydroxyl and carbonyl, and imine and amine groups with oxygen and nitrogen centers are responsible for a stronger inhibition activity of folate in comparison with two others.

Adewuyi et al. [26] are used *Khaya senegalensis* fatty hydroxylamide (KSFA) for inhibition of aluminium in 0.5 M HCl. The authors started from un-utilized seed oil, and in more steps, including esterification, transesterification, hydroxylation and amidation they synthetized KSFA. This compound with concentration of 0.001 mg/L exhibited 90.43 % of efficiency with concentration of 0.001mg/L, which is also related with the presence of heteroatoms in the KSFA.

Besides them, the agro-industrial waste based on rice bran oil is used to prepare a green corrosion inhibitor for the 1018 steel surface [27]. The authors showed that the addition of 10 and 25 ppm of inhibitor improves the inhibition of steel above 99 %. Fiori-Bimbi et al. [28] are used pectin from citrus peel as corrosion inhibitor of mild steel in HCl solutions. They have shown that pectin with concentration of 2.0 g/L assured efficient inhibition (94.2 %) of steel, at 45 °C by cathodic and anodic protection of metallic surface. Furthermore, the works of Ninčević Grassino et al. [29,30] showed that pectin isolated from tomato peel waste generated by canning industry act as eco-friendly corrosion inhibitor of tin. Corrosion tests performed in a NaCl/acetic acid/citric acid, at 25 °C pointed out that pectin is an efficient inhibitor (73 and 65.8 % for potentiodynamic polarization and electrochemical impedance spectroscopy), even at very low concentrations of 4 g/L.

Overall, the isolation of valuable-added compounds from abundant and under-utilized resources opens opportunity for production of new classes of inhibitors, which could be utilized as anticorrosion substances after their complete qualitative and quantitative analyses.

Table 3.   *Green corrosion inhibitors from biomass waste.*

Waste source	Main components	Material	Medium	Inhibitor concentration	Highest inhibition efficiency (%)	Refs.
Honeycomb	Quercetin	Stainless steel 304	0.5 M H2SO4	2000 mg/L	97.29	[15]
Egg shell	n.d.	Stainless steel 316	0.5 M H2SO4	2 - 10 g	94.74	[16]
Avocado oil	N, N-Diethylaminedialkyldiamide	API X-52 pipeline steel	3.5 % NaCl	10 ppm	94	[18]
Pinus radiata bark	Catechin, taxifolin	Steel	3.5 % NaCl (*w/v*)	n.r.	n.r.	[17]
Rice brain oil	Imidazoline		3.5 % NaCl	10 ppm	99.69	[27]
				25 ppm	99.65	
Lychee`s peel and seed	Epicatechin, epigallocatechin, cyanidin-3-glucoside	Mild steel	0.5 M HCl	600 mg/L	97.95	[19]
Prunus dulcis	Isorhamnetin-3-O-rutinoside, catechin, chlrogenic acid		0.1 M HCl	11.80 mg/L	93	[20]
Musa paradisa peel	Gallocatechin		1 M HCl	400 mg/L	91	[21]
			0.5 M H2SO4		85	
Citrus peel	Pectin		1 M HCl	2 g/L	94.2	[28]
Spirulina maxima	n.d.	Carbon steel	1 M HCl	100 m/L	96.40	[22]
Shrimp shell	Chitosan		1 M HCl	10^{-5} M	88.50	[23]
Sludge	n.d.		3.64 % NaCl	1000 mg/L	78.89	[24]

Gooseberry fruit shell	Sucrose, maltose, folate		1 M HCl	1000 ppm	91.9	[25]
Khaya senegalensis seed	Fatty acids (C18:1)	Aluminium	0.5 M HCl	0.001 mg/L	90.43	[26]
Tomato peel	Pectin	Tin	2 % NaCl%, 1 % acetic acid and 0.5 % citric acid	4 g/L	72.98	[29]
Tomato peel	Pectin		2 % NaCl%, 1 % acetic acid and 0.5 % citric acid	4 g/L	65.8	[30]

n.d. = not detected; n.r. = not reported

References

[1] G. Trabanelli, V. Carassiti, Mechanism and phenomenology of organic inhibitors, in: M.G. Fontaine, R.W. Stachle (Eds.), Advances in Corrosion Science and Technology, Plenum Press, New York, 1970, pp. 147-20. https://doi.org/10.1007/978-1-4615-8252-6_3

[2] P.B. Raja, M.S. Sethuraman, Matural products as corrosion inhibitor for metals in corrosive media - A review, Mater. Lett. 62 (2008) 113-116. https://doi.org/10.1016/j.matlet.2007.04.079

[3] J. Aljourani, K. Raeissi, M.A. Golozar, Benzimidazole and its derivatives as corrosion inhibitors for mild steel in 1 M HCl solution, Corros. Sci. 51 (2009) 1836-1843. https://doi.org/10.1016/j.corsci.2009.05.011

[4] S. Marzorati, L. Verotta, S.P. Trasatti, Green corrosion inhibitors from natural sources and biomass wastes, Molecules. 48 (2019) 1-24. https://doi.org/10.3390/molecules24010048

[5] A. Ninčević Grassino, Plant extracts as a natural corrosion inhibitors of metals and its alloys used in food preserving industry, in A. Méndez-Vilas (Ed.), Science within Food: Up to date Advances on Research and Education Ideas, Food Science Book Series 1, Formatex Research Center, Badajoz, 2017, pp. 185-193.

Materials Research Forum LLC
https://doi.org/10.21741/9781644901496-3

[6]   L.T. Popoola, Organic green corrosion inhibitors (OGCIs): a critical review, Corros. Rev. 37 (2019) 71-102. https://doi.org/10.1515/corrrev-2018-0058

[7]   S.A. Umoren, M.M. Solomon, I.B. Obot, R.K. Suleiman, A critical review on the recent studies on plant biomaterials as corrosion inhibitors for industrial metals, J. Ind. Eng. Chem. 76 (2019) 91-115. https://doi.org/10.1016/j.jiec.2019.03.057

[8]   C. Verma, E.E. Ebenso, I. Bahadur, M.A. Quraishi, An overview on plant extracts as environmental sustainable and green corrosion inhibitors for metals and alloys in aggressive corrosive media, J. Mol. Liq. 266 (2018) 577-590. https://doi.org/10.1016/j.molliq.2018.06.110

[9]   L.C. Pirvu, Polyphenols and herbal-based extracts at the basis of new antioxidant, Material Protecting Products, in: M. Aliofkhazraei (Ed.), Developments in Corrosion Protection, Intech Open, London, 2014, pp. 181-198.

[10] A. Dehghani, G. Bahlakeha, B. Ramezanzadeh, Green Eucalyptus leaf extract: A potent source of bio-active corrosion inhibitors for mild steel, Bioelectrochemistry. 130 (2019) 107339. https://doi.org/10.1016/j.bioelechem.2019.107339

[11] A. Ostovari, S.M. Hoseinieh, M. Peikari, S.R. Shadizadeh, S.J. Hashemi, Corrosion inhibition of mild steel in 1M HCl solution by henna extract: a comparative study of the inhibition by henna and its constituents (Lawsone, Gallic acid, α-d-Glucose and Tannic acid), Corros. Sci. 51 (2009) 1935-1949. https://doi.org/10.1016/j.corsci.2009.05.024

[12] M. Chevalier, F. Robert, N. Amusant, M. Traisnel, C. Roos, M. Lebrini, Enhanced corrosion resistance of mild steel in 1 M hydrochloric acid solution by alkaloids extract from *Aniba rosaeodora* plant: Electrochemical, phytochemical and XPS studies, Electrochim. acta 131 (2014) 96-105. https://doi.org/10.1016/j.electacta.2013.12.023

[13] P.C. Okafor, V.E. Ebiekpe, C.F. Azike, G.E. Egbung, E.A. Brisibe, E.E. Ebenso, Inhibitory action of *Artemisia annua* extracts and Artemisinin on the corrosion of mild steel in H_2SO_4 Solution, Int. J. Corros. 2012 (2011) 1-8. https://doi.org/10.1155/2012/768729

[14] A. Saxena, D. Prasad, R. Haldhar, G. Singh, A. Kumar, Use of *Saraca ashoka* extract as green corrosion inhibitor for mild steel in 0.5 M H_2SO_4, J. Mol. Liq. 258 (2018) 89-97. https://doi.org/10.1016/j.molliq.2018.02.104

[15] F. Gapsari, K.A. Madurani, F.M. Simanjuntak, A. Andoko, H. Wijaya, F. Kurniawan, Corrosion inhibition of honeycomb waste extracts for 304 stainless steel in sulfuric acid solution, Materials. 12 (2019) 1-15. https://doi.org/10.3390/ma12132120

[16] O. Sanni, A.P.I. Popoola, O.S.I. Fayomi, Enhanced corrosion resistance of stainless steel type 316 in sulphuric acid solution using eco-friendly waste product, Results Phys. 9 (2018) 225-230. https://doi.org/10.1016/j.rinp.2018.02.001

[17] L.F. Montoyaa, D. Contreras, A.F. Jaramillo, C. Carrascoc, K. Fernándezd, B. Schwederski, D. Rojas, M.F. Melendrez, Study of anticorrosive coatings based on high and low molecular weight polyphenols extracted from the Pine radiata bark, Prog. Org Coat. 127 (2019) 100-109. https://doi.org/10.1016/j.porgcoat.2018.11.010

[18] A. Cruz-Zabaleguia, E. Vazquez-Velezb, G. Galicia-Aguilara, M. Casales-Diazb, R. Lopez-Sesenesc, J.G. Gonzalez-Rodriguezd, L. Martinez-Gomez, Use of a non-ionic gemini-surfactant synthesized from the wasted avocado oil as a CO2- corrosion inhibitor for X-52 steel, Ind. Crops Prod. 133 (2019) 203-211. https://doi.org/10.1016/j.indcrop.2019.03.011

[19] L.L Liao, S. Mo, H.Q. Luo, N.B. Li, Corrosion protection for mild steel by extract from the waste of lychee fruit in HCl solution: experimental and theoretical studies, J. Colloid. Interf. Sci. 520 (2018) 41-49. https://doi.org/10.1016/j.jcis.2018.02.071

[20] S. Pal, H. Lgaz, P. Tiwari, I.M. Chung, G. Ji, R. Prakash, Experimental and theoretical investigation of aqueous and methanolic extracts of Prunus dulcis peels as green corrosion inhibitors of mild steel in aggressive chloride media, J. Mol. Liq. 276 (2019) 347-361. https://doi.org/10.1016/j.molliq.2018.11.099

[21] P. Tiwari, M. Srivastava, R. Mishra, G. Ji, R. Prakash, Economic use of Waste *Musa Paradisica* peels for effective control of mild steel loss in aggressive acid solutions, 6 (2018) 4773-4783. https://doi.org/10.1016/j.jece.2018.07.016

[22] L. Soares Rodrigues, A. Ferreira do Valle, E. D'Elia, Biomass of microalgae *spirulina maxima* as a corrosion inhibitor for 1020 carbon steel in acidic solution, Int. J. Electrochem. Sci. 13 (2018) 6169-6189. https://doi.org/10.20964/2018.07.11

[23] M.H.M. Hussein, M.F.El-Hady, H.A.H. Shehata, M.A. Hegazy, H.H.H. Hefni, Preparation of some eco-friendly corrosion inhibitors having antibacterial activity from sea food waste, J. Surfact. Deterg. 16 (2013) 233-242. https://doi.org/10.1007/s11743-012-1395-3

[24] L.C. Go, W. Holmes, D. Depan, R. Hernandez, Evaluation of extracellular polymeric substances extracted from waste activated sludge as a renewable corrosion inhibitor, Peer. J. (2019) e7193. https://doi.org/10.7717/peerj.7193

[25] A. Dehghani, G. Bahlakeha, B. Ramezanzadeh, A detailed electrochemical/theoretical exploration of the aqueous Chinese gooseberry fruit shell extract as a green and cheap corrosion inhibitor for mild steel in acidic solution, J. Mol. Liq. 282 (2019) 366-384. https://doi.org/10.1016/j.molliq.2019.03.011

[26] A. Adewuyi, R.A. Oderinde, Synthesis of hydroxylated fatty amide from underutilized seed oil of Khaya senegalensis: a potential green inhibitor of corrosion in aluminium, J. Anal. Sci. Technol. 9 (2018) 1-26. https://doi.org/10.1186/s40543-018-0158-9

[27] G. Salinas-Solanoa, J. Porcayo-Calderon, L.M. Martinez de la Escalera, J. Canto, M. Casales-Diaz, O. Sotelo-Mazon, John Henao, L. Martinez-Gomeza, Development and evaluation of a green corrosion inhibitor based on rice bran oil obtained from agro-industrial waste, Ind. Crops Prod. 119 (2018) 111-124. https://doi.org/10.1016/j.indcrop.2018.04.009

[28] M. Fioro-Bimbi, P.E. Alvarez, H. Vaca, C.A. Gervasi, Corrosion inhibition of mild steel in HCl solution by pectin, Corros. Sci. 92 (2015) 192-199. https://doi.org/10.1016/j.corsci.2014.12.002

[29] A. Ninčević Grassino, J. Halambek, S. Djaković, S. Rimac Brnčić, M. Dent, Z. Grabarić, Utilization of tomato peel waste from canning factory as a potential source for pectin production and application as tin corrosion inhibitor, Food Hydrocoll. 52 (2016) 265-274. https://doi.org/10.1016/j.foodhyd.2015.06.020

[30] A. Ninčević Grassino, S. Djaković, T. Bosiljkov, J. Halambek, Z. Zorić, V. Dragović-Uzelac, M. Petrović, S. Rimac Brnčić, Valorisation of tomato peel waste as a sustainable source for pectin, polyphenols and fatty acids recovery using sequential extraction, Waste and Biomass Valorization, 4 (2019). https://doi.org/10.1007/s12649-019-00814-7

[31] A.S. Abbas, É. Fazakas and T.I. Török, Corrosion studies of steel rebar samples in neutral sodium chloride solution also in the presence of a bio-based (green) inhibitor, Int. J. Corros. Scale Inhib. 7 (2018) 38-47.

[32] A. Marciales, T. Haile, B. Ahvazi, T.D. Ngo, J. Wolodko, Performance of green corrosion inhibitors from biomass in acidic media, Corros. Rev. 36 (2018) 239-266. https://doi.org/10.1515/corrrev-2017-0094

[33] K.E. Heusler, D. Landolt, S. Trasatti, Electrochemical corrosion nomenclature, Pure Appl. Chem. 61 (1989) 19-22. https://doi.org/10.1351/pac198961010019

[34] N. Perez, Electrochemistry and Corrosion Science, Kluwer Academic Publisher, Boston, 2004. https://doi.org/10.1007/b118420

[35] B.N. Popov, Corrosion inhibitors, in: B.N. Popov (Ed.), Corossion Engineering, Principles and Solved Problems, Elsevier, Amsterdam, 2015, pp. 581-597. https://doi.org/10.1016/B978-0-444-62722-3.00014-8

[36] O. Gharbi, S. Thomas, C. Smith, N. Birbilis, Chromate replacement: what does the future hold?, Mater. Degrad. 12 (2018). https://doi.org/10.1038/s41529-018-0034-5

[37] A. Režek Jambrak, Non-thermal and innovative processing technologies, in: P. Ferranti, E.M. Berry, J.R. Anderson (Eds.), Encyclopedia of Food Security and Sustainability, Elsevier, 2019, pp. 477-483. https://doi.org/10.1016/B978-0-08-100596-5.22285-3

[38] J. Azmir, I.S.M. Zaidul, M.M. Rahman, K.M. Sharif, A. Mohamed, F. Sahena, M.H.A. Jahurul, K. Ghafoor, N.A.N. Norulaini, A.K.M. Omar, Techniques for extraction of bioactive compounds from plant materials: A review, J. Food Eng. 117 (2013) 426-436. https://doi.org/10.1016/j.jfoodeng.2013.01.014

[39] D.C. Costa, H. Costa, T.G. Albuquerque, F. Ramos, M.C. Castilho, A. Sanches-Silva, Advances in phenolic compounds analysis of aromatic plants and their potential applications, Trends Food Sci. Tech. 45 (2015) 336-354. https://doi.org/10.1016/j.tifs.2015.06.009

[40] K. Xhanari, M. Fingšar, M. Knez-Hrnčić, U. Maver, Ž. Knez, B. Seiti, Green corrosion inhibitors for aluminium and its alloys: a review, RSC Adv. 7 (2017) 27299-27330. https://doi.org/10.1039/C7RA03944A

[41] M. Srivastava, P. Tiwari, S.K. Srivastava, A. Kumar, G. Ji, R. Prakash, Low cost aqueous extract of *Pisum sativum* peels for inhibition of mild steel corrosion, J. Mol. Liq. 254 (2018) 357-368. https://doi.org/10.1016/j.molliq.2018.01.137

[42] S. Burt, Essential oils: their antimicrobial properties and potential application in foods-A review, Inter. J. Food Microb. 94 (2004) 223-253. https://doi.org/10.1016/j.ijfoodmicro.2004.03.022

[43] L. Afia, R. Salghi, L. Bammou, El. Bazzi, B. Hammouti, L. Bazzi, A. Bouyanzer, Anti-corrosive properties of Argan oil on C38 steel in molar HCl solution, J. Saudi Chem. Soc. 18 (2014) 19-25. https://doi.org/10.1016/j.jscs.2011.05.008

[44] D.B. Hmamou, R. Salghi, A. Zarrouk, H. Zarrouk, M. Errami, B. Hammouti, L. Afia, L. Bazzi, L. Bazzi, Adsorption and corrosion inhibition of mild steel in hydrochloric acid solution by verbena essential oil, Res. Chem. Intermed. 39 (2013) 973-989. https://doi.org/10.1007/s11164-012-0609-7

[45] D.B. Hmamou, R. Salghi, A. Zarrouk, O. Benali, F. Fade, H. Zarrok, B. Hammouti, Carob seed oil: an efficient inhibitor of C38 steel corrosion in hydrochloric acid, Int. J. Ind. Chem. 3 (2012) 1-9. https://doi.org/10.1186/2228-5547-3-25

[46] M. Bendahou, M. Benabdellah, B. Hammouti, A study of rosemary oil as a green corrosion inhibitor for steel in 2 M H_3PO_4, Pigm. Resin Technol, 35 (2006) 95-100. https://doi.org/10.1108/03699420610652386

[47] D. Ben Hmamou, R. Salghi, A. Zarrouk, B. Hammouti, S.S. Al-Deyab, L. Bazzi, H. Zarrok, A. Chakir, L. Bammou, Corrosion inhibition of steel in 1 m hydrochloric acid medium by chamomile essential oils, Int. J. Electrochem. Sci. 7 (2012) 2361-2373.

[48] N. Lahhit, A. Bouyanzer, J.M. Desjobert, B. Hammouti, R. Salghi, J. Costa, C. Jama, F. Bentiss, L. Majidi, Fennel (Foeniculum Vulgare) Essential oil as green corrosion inhibitor of carbon steel in hydrochloric acid solution, Port. Electrochim. Acta. 29 (2011) 127-138. https://doi.org/10.4152/pea.201102127

[49] L. Afia, O. Benali, R. Salghi, E.E. Ebenso, S. Jodeh, M. Zougagh, B. Hammouti, Steel Corrosion inhibition by acid garlic essential oil as a green corrosion inhibitor and sorption behavior, Int. J. Electrochem. Sci. 9 (2014) 8392-8406.

[50] M. Chraibi, K. Fikri Benbrahim, H. Elmsellem, A. Farah, I. Abdel-Rahman, B. El Mahi, Y. Filali Baba, Y. Kandri Rodi, F. Hlimi, Antibacterial activity and corrosion inhibition of mild steel in 1.0 M hydrochloric acid solution by M. piperita and M. pulegium essential oils, J. Mater. Environ. Sci. 8 (2017) 972-981.

[51] Y. El Ouadi, A. Bouyanzer, L. Majidi, J. Paolini, J.M. Desjobert, J. Costa, A. Chetouani, B. Hammouti, S. Jodeh, I. Warad, Y. Mabkhot, T. Ben Hadda, Evaluation of Pelargonium extract and oil as eco-friendly corrosion inhibitor for steel in acidic chloride solutions and pharmacological properties, Res. Chem. Intermed. 41 (2015) 7125-7149. https://doi.org/10.1007/s11164-014-1802-7

[52] J. Halambek, M. Cvjetko Bubalo, I. Radojčić Redovniković, K. Berković, Corrosion behaviour of aluminium and AA5754 alloy in 1% acetic acid solution in presence of laurel oil, Int. J. Electrochem. Sci. 9 (2014) 5496-5506. https://doi.org/10.1016/j.ecoenv.2013.10.019

[53] J. Halambek, A. Žutinić, K. Berković, Ocimum basilicumL. oil as corrosion inhibitor for aluminium in hydrochloric acid solution, Int. J. Electrochem. Sci. 8 (2013) 11201-11214.

[54] J. Halambek, K. Berković, Inhibitive action of Anethum graveolens L. oil on aluminium corrosion in acidic media, Int. J. Electrochem. Sci. 7 (2012) 8356-8368.

[55] K. Dahmani, M. Galai, M. Cherkaoui, A. El Hasnaoui, A. El Hessni, Cinnamon essential oil as a novel eco-friendly corrosion inhibitor of copper in 0.5 M Sulfuric Acid medium, J. Mater. Environ. Sci. 8 (2017) 1676-1689.

Materials Research Forum LLC
https://doi.org/10.21741/9781644901496-3

[56] L. Bammou, M. Mihit, R. Salghi, A. Bouyanzer, S.S. Al-Deyab, L. Bazzi, B. Hammouti, Inhibition effect of natural artemisia oils towards tinplate corrosion in HCl solution: Chemical characterization and electrochemical study, Int. J. Electrochem. Sci.6 (2011) 1454-1467.

[57] L. Bammou , B. Chebli , R. Salghi , L. Bazzi , B. Hammouti , M. Mihit, H. Idrissi, Thermodynamic properties of Thymus satureioides essential oils as corrosion inhibitor of tinplate in 0.5 M HCl: chemical characterization and electrochemical study, Green Chem. Lett. Rev. 3 (2010) 173-178. https://doi.org/10.1080/17518251003660121

[58] J. Halambek,K. Berković, J. Vorkapić-Furač, The influence of Lavandula angustifolia L. oil on corrosion of Al-3Mg alloy, Corros. Sci. 52 (2010) 3978-3983. https://doi.org/10.1016/j.corsci.2010.08.012

[59] J. Halambek, K. Berković, J. Vorkapić-Furač, Laurus nobilis L. oil as green corrosion inhibitor for aluminium and AA5754 aluminium alloy in 3 % NaCl solution, Mater. Chem. Phys. 137 (2013) 788-795. https://doi.org/10.1016/j.matchemphys.2012.09.066

[60] R. Haldhar, D. Prasad, A. Saxena, Armoracia rusticana as sustainable and eco-friendly corrosion inhibitor for mild steel in 0.5M sulphuric acid: Experimental and theoretical investigations, J. Environ. Chem. Eng. 6 (2018) 5230-5238. https://doi.org/10.1016/j.jece.2018.08.025

[61] P. C. Okafor, E. E. Ebenso, U.J. Ekpe, Azadirachta Indica Extracts as corrosion inhibitor for mild steel in acid medium, Int. J. Electrochem. Sci. 5 (2010) 978-993.

[62] X. Li, S. Deng, H. Fu, Inhibition of the corrosion of steel in HCl, H_2SO_4 solutions by bamboo leaf extract, Corros. Sci. 62 (2012) 163-175. https://doi.org/10.1016/j.corsci.2012.05.008

[63] S. Deng, X. Li, Inhibition by Ginkgo leaves extract of the corrosion of steel in HCl and H_2SO_4 solutions, Corros. Sci. 55 (2012) 407-415. https://doi.org/10.1016/j.corsci.2011.11.005

[64] I.B. Obot, N.O. Obi-Egbedi, Ginseng Root: A new efficient and effective eco-friendly corrosion inhibitor for aluminium alloy of type aa 1060 in hydrochloric acid solution, Int. J. Electrochem. Sci. 4 (2009) 1277-1288.

[65] M. Dahmani, A. Et-Touhami, S.S. Al-Deyab, B. Hammouti, A. Bouyanzer, Corrosion inhibition of C38 steel in 1 M HCl: A comparative study of black pepper extract and its isolated piperine, Int. J. Electrochem. Sci. 5 (2010) 1060-1069.

[66] A. Dehghani , G. Bahlakeha, B. Ramezanzadehb, M. Ramezanzadeh, Potential of Borage flower aqueous extract as an environmentally sustainable corrosion inhibitor

Materials Research Foundations**107** (2021) 46-69   https://doi.org/10.21741/9781644901496-3

for acid corrosion of mild steel: Electrochemical and theoretical studies, J. Mol. Liq. 277 (2019) 895-911. https://doi.org/10.1016/j.molliq.2019.01.008

[67] M. Faustin, A. Maciuk, P. Salvin, C. Roos, M. Lebrini, Corrosion inhibition of C38 steel by alkaloids extract of Geissospermum laeve in 1 M hydrochloric acid: Electrochemical and phytochemical studies, Corros. Sci. 92 (2015) 287-300. https://doi.org/10.1016/j.corsci.2014.12.005

[68] I.B. Obot, E.E. Ebenso, Z.M. Gasem, Eco-friendly corrosion inhibitors: Adsorption and inhibitive action of ethanol extracts of Chlomolaena Odorata L. for the corrosion of mild steel in H_2SO_4 solutions, Int. J. Electrochem. Sci. 7 (2012) 1997-2008.

[69] E.A. Noor, Potential of aqueous extract of Hibiscus sabdariffa leaves for inhibiting the corrosion of aluminum in alkaline solutions, J. Appl. Electrochem. 39 (2009) 1465-1475. https://doi.org/10.1007/s10800-009-9826-1

[70] A. Saxena, D. Prasad, R. Haldhar, G. Singh, A. Kumar, Use of Saraca ashoka extract as green corrosion inhibitor for mild steel in 0.5 M H_2SO_4, J. Mol. Liq. 258 (2018) 89-97. https://doi.org/10.1016/j.molliq.2018.02.104

[71] X.H. Li, S.H. Deng, H. Fu, Inhibition by Jasminum nudiflorum Lindl. leaves extract of the corrosion of cold rolled steel in hydrochloric acid solution, J. Appl. Electrochem. 40 (2010) 1641-1649. https://doi.org/10.1007/s10800-010-0151-5

[72] Q. Wang, B. Tan, H. Bao, Y. Xie, Y. Moua, P. Li, D. Chena, Y. Shi, X. Li, W. Yang, Evaluation of Ficus tikoua leaves extract as an eco-friendly corrosion inhibitor for carbon steel in HCl media, Bioelectrochemistry. 128 (2019) 49-55. https://doi.org/10.1016/j.bioelechem.2019.03.001

[73] A. Singh, E.E. Ebenso, M. A. Quraishi, Theoretical and electrochemical studies of cuminum cyminum (jeera) extract as green corrosion inhibitor for mild steel in hydrochloric acid solution, Int. J. Electrochem. Sci. 7 (2012) 8543-8559.

[74] Z. Sanaeia, M. Ramezanzadeha, G. Bahlakehb, B. Ramezanzadeha, Use of Rosa canina fruit extract as a green corrosion inhibitor for mild steel in 1 M HCl solution: A complementary experimental, molecular dynamics and quantum mechanics investigation, J. Ind. Eng. Chem. 69 (2019) 18-31. https://doi.org/10.1016/j.jiec.2018.09.013

[75] T. H. Ibrahim, M. A. Zour, Corrosion Inhibition of mild steel using fig leaves extract in hydrochloric acid solution, Int. J. Electrochem. Sci. 6 (2011) 6442-6455.

Sustainable Corrosion Inhibitors
Materials Research Foundations**107** (2021) 70-100

Materials Research Forum LLC
https://doi.org/10.21741/9781644901496-4

Chapter 4

Polysaccharide as Green Corrosion Inhibitor

Y. Dewangan, A.K. Dewangan, D.K. Verma*

Department of Chemistry, Government Digvijay Autonomous Postgraduate College,
Rajnandgaon, Chhattisgarh, INDIA, 491441

*dakeshwarverma@gmil.com

Abstract

The carbohydrates associated with polysaccharide glycosidic bonds are tightly chained, usually linear and highly branched complex molecules. Their structure mainly consists of hydroxyl groups in the form of functional groups, in which an oxygen heterogeneous atom is present. Some polysaccharides have hetero atoms. Nitrogen and Sulfur in addition to oxygen, which have unshared electron pairs. Hetero atoms easily share their electron pair to the vacant d orbitals of the metal ion and prevent the metal from corrosion. Polysaccharides are biodegradable, renewable, inexpensive and environment friendly due to which they are easily used as corrosion inhibitors. The present study mentions some major research work in which polysaccharides are used as corrosion inhibitors. Their mixed type nature has been reported in most research papers, and in the case of steel metal, they mainly follow the Langmuir adsorption isotherm. Chemical (gravimetric analysis) and electrochemical (EIS & PDP) studies are frequently used for the corrosion inhibition study. Some of the current research papers have also used computational or theoretical studies such as quantum chemical study and MD simulation. At the end of this book chapter, a discussion is also given regarding further research and direction related to the topic.

Keywords

Polysaccharide, Heteroatom, Metal, Corrosion Inhibitor, Electron Impedance Spectroscopy, Density Function Theory

Contents

1.   Introduction

Due to the usefulness of metals and alloys, it is very important to research the cause of corrosion and their inhibition. Commonly HCl and H_2SO_4 Corrosive Media are used to study Corrosion Behavior in metals and alloys. During Acid pickling, Chemical cleaning and Rust cleaning, these acids are used frequently which causes loss of metals [1-3]. Corrosion inhibitors are used to prevent this loss of metals, even small amounts of which are useful in preventing metal corrosion, thereby prolonging the lifespan of metals. Most reported corrosion-resistant organic compounds are expensive, hazardous, and toxic in nature from manufacture to use, due to which continuous research is underway to reduce their use to a great extent [4]. In this direction, using natural corrosion inhibitors is an inexpensive and simple method, which is biodegradable, cheap and environment friendly. There is some old research in which various scientists and engineers used the products derived from the plant such as root, fruit, flower, stem, gum, and water as green corrosion inhibitors [5-11]. Polysaccharides are natural polymers that are environmentally friendly, biodegradable, inexpensive and stable due to which they are used as metal corrosion inhibitors. These are inexpensive as well as renewable and are readily available which contain OH groups, hetero atoms, and alkyl chain lengths, due to which they are easily deposited onto the metal surface. Some of the major works mentioned in this book chapter are used as corrosion inhibitors [11-15]. From the above research, it can be concluded that plant extract usually contains complex organic compounds like amino acids, proteins, alkaloids, tannins, and carbohydrates [16]. These biomolecules have polar functional groups, which are mainly hetero atoms such as O, N, P, S and polygons [17]. Their structure reflects physiochemical adhesion between the organic molecule and the metal surface, which is a kind of donor acceptor property. Due to such a structure, they are easily absorbed on the metal surface, and form protective layers that block the active site of the metal surface for further corrosion. EIS and PDP are extensively used for the electrochemical analysis for inhibition determination [18,19]. Similarly, SEM-EDS applied for modification seen on the metal surface. DFT calculation and MD simulation are extensively used as theoretical calculation [20-23]. Fig. 1 represents the general characteristics of corrosion inhibitor.

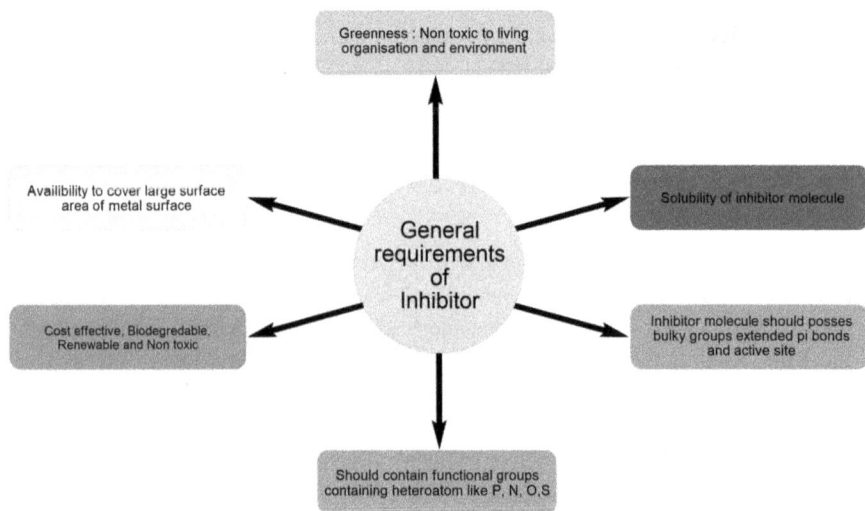

Figure 1.   General characteristics of corrosion inhibitor.

2.   Polysaccharides as prominent corrosion inhibitor

2.1   Corrosion inhibition of ferrous materials

HPMC has been applied as potential corrosion inhibitor for MS corrosion in aggressive solution. HPMC follows Freundlich adsorption isotherm and its mixed type nature has been reported. According to EIS analysis HPMC exhibited 91.18 % I when applied with KI i.e. synergistic effect [24]. Guar gum applied as potential corrosion inhibitor for carbon steel in corrosive solution (M. Abdullah 2004) and reported that it follows Langmuir adsorption isotherm. Weight loss study shown the surface coverage (θ) of 0.939 in 1500 ppm concentration of inhibitor [25]. M. Mobin et al. (2011) applied various techniques like WL, PDP, SEM, where weight loss revels 67.41 % inhibition efficiency in 500 ppm concentration at 30°C of starch and surfactants. ΔG adsorption value -35.54 KJMol^{-1} observed at 30°C when starch applied with CTAB [26]. Synergistic effect of Gum Acacia and surfactants namely CTAB and SDBA used as corrosion inhibitor for mid steel in corrosive solution (0.1 M H_2SO_4). Analysis performed at temperature ranging from 30°C to 60°C [27]. Synergistic effect of Xanthan gum with surfactants was applied as corrosion inhibitor. 74.24 % ion-exchange efficiency (*IE*) obtained with XG at 1000 ppm at 30°C. Techniques such as UV-VIS, SEM, EIS and DFT calculation are applied for inhibition analysis [28]. UV-VIS, AFM, SEM, FTIR and EIS techniques are applied to check the efficiency of Plantago (polysaccharide) for carbon steel corrosion in acidic medium (1M HCl). -250 to +250 mV applied for electrochemical analysis (PDP) and found 86.11% *IE* at 400 ppm concentration [29]. EIS, SEM, PDP, WL analysis revels that mixed type nature of pectin for MS in corrosive solution. Corrosion rate of 0.062 mg cm^{-2}h^{-2} and inhibition efficiency of 90.3 are obtained at 2 g/L amount of inhibitor [30]. Analysis suggest the Dubinin-radushkevich and Langmuir adsorption isotherm of CMC at 30°C to 60°C and 64.8 inhibition efficiency as per weight loss analysis [31]. Reported methods suggest the potential property of HEC towards carbon steel and follows the Langmuir model [32]. Modified Cassava Starches were used as potential inhibitor for carbon steel corrosion and proved as efficient inhibitor for MS corrosion in 200 mg L^{-1} NaCl [33]. EIS, PDP, FTIR, Gravimetric analysis reveals the corrosion rate of 0.06 mg cm^{-2} h^{-1} (30°C) and inhibition efficiency of 96.30 at 60°C at 400 ppm (30°C) [34]. For carbon steel corrosion, Hydroxyethyl Cellulose and surfactants applied in 1M HCl solution. Various corrosion monitoring techniques such as EIS, Quantum Chemical Analysis showed the good agreement for the potential of inhibitor [35]. Eddy et al (2011) showed using gravimetric analysis that the corrosion rate 0.000197 gcm^{-1} at 303 K at 0.5 gl^{-1} conc. of inhibitor [36]. They proposed the synergistic effect and found 6453% I in 0.5 g/L concentration at 300 K for MS corrosion by using (AL) anogessus Leocarpus gum [37]. Eddy (2012) and co-workers reported the Daniella Oliverri (DO) gum exudates as potential corrosion inhibitor on MS (0.1 M HCl) and found that do follows the Langmuir adsorption isotherm [38]. WL measurement suggest that inhibitor molecules follow the Langmuir and Freundlich model [39]. Natural polysaccharide namely Polyacrylamide compound with Okra mucilage exhibited optimum inhibition efficiency (96.6%) in 100 ppm concentration at room temperature [40]. Natural polysaccharide namely hydroxypropyl cellulose, glucose and Gellan gum evaluated as corrosion protection of cast iron. EIS shows that hydroxypropyl cellulose shows optimum % inhibition at room temperature [41]. PDP and WL analysis reveals the mixed nature of Guar gum at its optimum concentration [42]. Pipeline steel corrosion protection monitoring by techniques like PDP, FTIR, EIS and found mixed nature of inhibitor [43]. SEM analysis shows the protective layer of xanthum gum on a MS surface [44]. CMC proved as effective inhibitor for MS corrosion in electrolytic media as per WL, HE, TG analysis [45]. Synergistic effect between KX (KBr, KI, KCl) and CMC exhibited effective corrosion resistance [46]. SEM analysis shows the synergistic protective layer of CMC and zinc ion onto carbon steel [47]. EDS analysis shows the modified chemical composition onto the carbon steel surface [48]. DFT calculation applied to determine inhibition property of synergism (HEC & KI) [49]. TS and DPMP act synergistically towards metal protection proved by FTIR and other analysis [50]. Surface analysis like SEM, XPS, FTIR, exhibited the barrier on metal surface of zinc ion, phosphonic acid and pectin [51]. SEM, FTIR analysis shows the inhibitory action of pectin enhanced by using it with grafted polymer [52]. EIS analysis proved the chitosan as efficient inhibitor for MS in corrosive media [53]. ATC based conjugated polymer act as efficient mixed type towards MS corrosion [54]. Nanogels (chitosan grafted) act as mixed nature towards MS corrosion inhibitor [55]. Surface analysis like EDS, XRD shows the protective layer deposited onto MS surface [56]. CCBA and CMCUGA synergistic make chitosan as potential corrosion inhibitor [57]. Techniques like AFM, XRD, EDS, EIS applied OFC as corrosion inhibitor [58]. The molecular structure, ferrous metals and media, techniques applied, nature of inhibitors and outcomes of the research depicted in Table 1.

Table 1. The molecular structure, ferrous metals and media, techniques applied, nature of inhibitors and outcomes of the research.

SN	Structure of Inhibitor	Metal & Electrolytic Media and nature of inhibitor	Technique Used	Salient Features	Ref.
1	hydroxypropyl methylcellulose	MS/ 0.5 M H$_2$SO$_4$/ Mixed Type/ Freundlich model	WL, EIS, PDP, OCP, QCC, DFT	HPMC has been applied as potential corrosion inhibitor for MS corrosion in aggressive solution. HPMC follows Freundlich adsorption isotherm and its mixed type nature has been reported. According to EIS analysis HPMC exhibited 91.18 % *I* when applied with KI i.e. synergistic effect.	24

2	Guar gum	CS/ 1 M H$_2$SO$_4$/ Mixed Type/ Langmuir model	WL	Guar gum applied as potential corrosion inhibitor for CS in corrosive solution (M. Abdullah 2004) and reported that it follows Langmuir adsorption isotherm. WL shown the surface coverage (θ) of 0.939 in 1500 ppm concentration of inhibitor	25
3	Amylopectin	MS/ 0.1 M H$_2$SO$_4$/ Anodic/ Langmuir model	EIS, WL, PDP	M. Mobin et al (2011) applied various technique like WL, PDP, DEM, where weight loss revels the 67.41 % inhibition efficiency in 500 ppm concentration at 30°C of starch and surfactants. ΔG adsorption value -35.54 KJMol^{-1} observed at 30°C when starch applied with CTAB	26
4	Gum Acacia and surfactants namely CTAB and SDBS CTAB SDBS	MS/ 0.1 M H$_2$SO$_4$/ Mixed Type/ Freundlich adsorption isotherm	WL, SEM, AFM	Synergistic effect of Gum Acacia and surfactants namely CTAB and SDBA used as corrosion inhibitor for mid steel in corrosive solution (0.1 M H$_2$SO$_4$). Analysis performed at temperature ranging from 30°C to 60°C	27

5	Xanthan gum	MS/ 1 M HCl/ Mixed/ Langmuir model	WL, GA, PDP, EIS, QCC, SEM, UV-VS	Synergistic effect of Xanthan gum with surfactants was applied as corrosion inhibitor. 74.24 % IE obtained with XG at 1000 ppm at 30°C. Technique such as UV-VIS, SEM, EIS and DFT calculation applied for inhibition analysis	28
6	gum arabic	CS/ 1 M HCl/ Mixed/ Langmuir model	WL, SEM, AFM, UV-VIS, FTIR, EIS, GA, PDP	UV-VIS, AFM, SEM, FTIR and EIS technique applied to check efficiency of Plantago (polysaccharide) for carbon steel corrosion in acidic medium (1M HCl). -250 to +250 mV applied for electrochemical analysis (ppp) and found 86.11 IE at 400 ppm concentration	29
7	pectin	MS/ 0.5 M HCl/ Mixed Type	OCP, WL, SEM, PDP, EIS	EIS, SEM, PDP, WL analysis revels that mixed type nature of pectin for MS in corrosive solution. Corrosion rate of 0.062 $mgcm^{-2}h^{-2}$ and inhibition efficiency of 90.3 obtained at 2 g/L amount of inhibitor	30

8	carboxymethyl cellulose cellulose	MS/ 2 M H_2SO_4/ Mixed/ Langmuir model	WL	Analysis suggest the dubinin-radushkevich and Langmuir adsorption isotherm of CMC at 30°C to 60°C and 64.8 inhibition efficiency as per weight loss analysis	31
9	Hydroxyethylcellulose	Carbon Steel/ 3.5% NaCl/ Mixed Type/ Langmuir model	OCP, QCC, EIS, SEM, DFT, PDP	Reported methods suggest the potential property of HEC towards CS and follows Langmuir model	32
10	carboxymethylated starch cassava starch	CS/ 0.05 M HCl	WL, DFT, XPS, EIS	Modified Cassava Starches used as potential inhibitor for carbon steel corrosion and proved as efficient inhibitor for MS corrosion in 200 mg L^{-1} NaCl	33

11	chitosan arabinogalactan	CS/ 1 M HCl/ Mixed Type/ Langmuir model	WL, GA, SEM PDP, OCP, QCC, DFT, AFM, UV-VS, EIS, MCS, AFM	EIS, PDP, FTIR, Gravimetric analysis reveals the corrosion rate of 0.06 mg cm^{-2} h^{-1} (30°C) and inhibition efficiency of 96.30 at 60°C at 400 ppm (30°C)	34
12	Hydroxyethylcellulose	CS/ 1 M HCl/ Mixed Type/ Langmuir model	PDP, UV-VIS, WL, EDX, AFM, QCC, OCP, SEM, EIS	For carbon steel corrosion, Hydroxyethyl Cellulose and surfactants applied in 1M HCl solution. Various corrosion monitoring techniques such as EIS, Quantum Chemical Analysis shown the good agreement for the potential of inhibitor	35
13	Guar gum	MS/ 0.1 M HCl/ Mixed/ Langmuir model	WL, EIS, PDP	Gravimetric analysis shows the corrosion rate 0.000197 gcm^{-1} at 303 K at 0.5 gl^{-1} conc. Of inhibitor	36

14	 SUCROSE	MS/ 0.1 M HCl/ Langmuir model	GCMS, FTIR, WL	Eddy et al (2011) proposed the synergistic effect and found 6453% I in 0.5 g/L concentration at 300 K for MS corrosion by using (AL) anogessus Leocarpus gum	37
15	 SUCROSE	MS/ 0.1 M HCl/ Langmuir model	GA, GCMS, FTIR, WL	Eddy (2012) and coworkers reported the Daniella Oliverri (DO) gum exudates as potential corrosion inhibitor on MS (0.1 M HCl) and found that do follows the Langmuir adsorption isotherm	38
16	 Guar gum	MS/ 2 M HCl/ Langmuir and Freundlich model	GA, GM, TM, GCMS, WL	WL measurement suggest that inhibitor molecules follows Langmuir and Freundlich model	39

17	Structure of O-g-PAM. Okra mucilage	MS// 0.5 M H_2SO_4/ Cathodic/ Langmuir model	GA, WL, OCP, SEM, EIS, XRD, FTIR, PDP	Natural polysaccharide namely Polyacrylamide compound with Okra mucilage exhibited optimum inhibition efficiency (96.6%) in 100 ppm concentration at room temperature	40
18		Cast Iron/ 1 M HCl/ Mixed / Langmuir and Freundlich model	WL, PDP, OCP, SEM, XRD, EIS	Natural polysaccharide namely hydroxypropyl cellulose, glucose and Gellan gum evaluated as corrosion protection of cast iron. EIS shows that hydroxypropyl cellulose shows optimum % inhibition at room temperature	41

19	Guar gum gum arabic	CS (L-52 grade)/ 1 M H$_2$SO$_4$ containing NaCl/ Mixed nature	PDP, WL	PDP and WL analysis reveals the mixed nature of Guar gum at its optimum concentration	42
20	gum arabic	Pipeline steel (API 5L X42)/1 M HCl/ Mixed type	PDP, FTIR, EIS	Pipeline steel corrosion protection monitoring by techniques like PDP, FTIR, EIS and found mixed nature of inhibitor	43

21	Xanthan gum and Xanthan gum-polyacrylamide conjugate	MS/ 15% HCl/ Mixed type	PDP, SEM, EIS, WL	SEM analysis shows the protective layer of xanthum gum on MS surface	44
22	Carboxymethyl cellulose	MS/ 2 M H_2SO_4	WL, HE, TG	CMC proved as effective inhibitor for MS corrosion in electrolytic media as per WL, HE, TG analysis	45
23		MS (AISI 1005 grade)/ 2 M H_2SO_4	WL, HE	Synergistic effect between KX(KBr, KI, KCl) and CMC exhibited effective corrosion resistance	46
24	$CMC-Zn^{2+}$	Carbon steel/ Ground water (pH 11)	EIS, SEM, WL	SEM analysis shows the synergistic protective layer of CMC and zinc ion onto carbon steel	47

25	Hydroxyethylcellulose	CS (1018 grade)/ 3.5% NaCl/ Mixed type	SEM, DFT, EIS, EFM, PDP, EDS,	EDS analysis shows the modified chemical composition onto carbon steel surface	48
26	Hydroxyethylcellulose	MS/ 0.5 M H_2SO_4/ Mixed type	EIS, DFT, WL, PDP	DFT calculation applied to determine inhibition property of synergism (HEC & KI)	49
27	tapioca starch	MS/ 1 N HCl/ Mixed type	PDP, WL, EIS, FTIR	TS and DPMP act synergistically towards metal protection proved by FTIR and other analysis	50
28		Carbon steel 60 ppm Cl^-/ Mixed type	SEM, WL, XPS, EIS, FTIR	Surface analysis like SEM, XPS, FTIR, exhibited the barrier on metal surface of zinc ion, phosphonic acid and pectin	51
29		MS/ 3.5% NaCl/ Mixed (Cathodic mainly)	WL, EDS, TEM, TGA	SEM, FTIR analysis shows the inhibitory action of pectin enhanced by using it with grafted polymer	52
30	chitosan	MS/ 0.1 M HCl/ Mixed type	WL, SEM, PDP, EIS	EIS analysis proved the chitosan as efficient inhibitor for MS in corrosive media	53

31	chitosan	MS/ 0.5 M H_2SO_4/ Mixed type	PDP, SEM, EIS	ATC based conjugated polymer act as efficient mixed type towards MS corrosion	54
32	chitosan	MS/ 1 M HCl/ Mixed	PDP, EIS	Nanogels (chitosan grafted) act as mixed nature towards MS corrosion inhibitor	55
33	chitosan	MS/ 0.1 N HCl/ Cathodic type	XRD, PDP, SEM, FTIR, EIS, EDS	Surface analysis like EDS, XRD shows the protective layer deposited onto MS surface	56
34	chitosan	MS/ 2% NaCl	Fluidization	CCBA and CMCUGA synergistic make chitosan as potential corrosion inhibitor	57
35	chitosan	MS/ 1 M HCl/ Mixed type	AFM, PDP, XRD, EDS, EIS, SEM	Techniques like AFM, XRD, EDS, EIS applied OFC as corrosion inhibitor	58

2.2 Corrosion inhibition of non-ferrous materials

Gravimetric analysis on aluminium in aggressive media of xanthum gum suggest that it follows the Temkin model [59]. Corrosion inhibition on aluminium in H_2SO_4 medium was evaluated by using Ficustricopoda gum (FP gum) at 0.5 g/L concentration. FP gum shows optimum inhibition i.e. 85.80 % [60]. Ficusbenjamina (FB) gum used and exhibited surface coverage of 0.875 at 0.5 g/L concentration [61]. Gum Arabic based research work has been published for corrosion inhibitor for Al and found effective inhibition property [62]. 0.5 M H_2SO_4 corrosive solution applied for corrosion inhibition of aluminum by applying techniques such as DFT and WL [63]. Corrosion inhibition on AA6061 alloy studied using tapioca starch in natural saline water where EIS study reveals the 96.00% inhibition efficiency at 1000 ppm and ΔG 30.469 KJmol^{-1} value suggest the mixed type nature of the inhibitor [64]. Gum Arabic (GA) treated and tested as potential corrosion inhibitor on aluminium corrosion inhibition in 1 M NaOH by various parameters like hydrogen evolution, adsorption, kinetics, gravimetric etc. Thermometric analysis reveals the % inhibition of 75.86 at 0.5 g/L amount of inhibitor [65]. S.A. Umoren and coworkers (2008) reveals that dacroydesedulis exudates gum shows % inhibition of 42.0 at 5 g/L concentration (30 °C temperature) [66]. Researchers reported the CMC as mixed type inhibitor as per WL analysis [67]. Analysis like CV, EIS and WL applied for Al & Al-Si alloys corrosion at 0.1 M NaOH and find efficient towards corrosion inhibition [68]. Scientists proved PA as efficient inhibitor by applying HE and WL techniques [69]. WL measurement suggest the good inhibition property of PA which follows Langmuir model [70]. Chitosan studied by applying PDP, DFT, WL, EFM, EIS techniques as Cu corrosion inhibitor and reported as mixed type nature [71]. Various techniques like EIS, FTIR, TGA, SEM, PDP were applied and reported the mixed type behaviour of chitosan [72]. Various techniques applied and result reveals the mixed type inhibition through the polysaccharides [73]. The molecular structure, ferrous metals and media, techniques applied, nature of inhibitors and outcomes of the research summarized in Table 2.

Table 2. The molecular structure, non-ferrous metals and media, techniques applied, nature of inhibitors and outcomes of the research.

SN	Structure of Inhibitor	Metal, Electrolytic Media and nature of inhibitor	Technique Used	Outcomes	Ref.
1	Xanthum gum	Aluminium/ 0.5 M HCl/ Mixed Type/ Temkin model	WL, PDP, OCP, ECP	Gravimetric analysis on Al in aggressive media of xanthum gum suggest that it follows the Temkin model	59

2	SUCROSE	Aluminum/ 0.01 M H_2SO_4 / Langmuir model	WL, GCMS	Corrosion inhibition on aluminum in H_2SO_4 medium was evaluated by using Ficustricopoda gum (FP gum) at 0.5 g/L concentration. FP gum shows optimum inhibition i.e. 85.80.	60
3	SUCROSE	Aluminum/ 0.1 M H_2SO_4/ FDR adsorption model	WL, GCMS, FTIR	Ficusbenjamina (FB) gum used and exhibited surface coverage of 0.875 at 0.5 g/L concentration	61
4	gum arabic	Aluminium/ 0.1 M HCl/ Langmuir model	WL, GA, QCC, FTIR, GCMS	Gum Arabic based research work has been published for corrosion inhibitor for Al and found effective inhibition property	62
5	hydroxypropyl methylcellulose	Aluminium/ 0.5 M H_2SO_4/ Cathodic/ Freundlich model	WL, EIS, QCC, DFT, PDP, OCP	0.5 M H_2SO_4 corrosive solution applied for corrosion inhibition of aluminum by applying techniques such as DFT and WL	63

Materials Research Forum LLC

https://doi.org/10.21741/9781644901496-4

| 6 | | Aluminium and its Alloy/ HCl/ Mixed Type/ Langmuir model | WL, GA, PDP, LPR, EIS, SEM, EDS, OCP, FTIR | Corrosion inhibition on AA6061 alloy studied using tapioca starch in natural saline water where EIS study reveals the 96.00% inhibition efficiency at 1000 ppm and Δ G 30.469 KJmol^{-1} value suggest the mixed type nature of inhibitor | 64 |
| 7 | | Aluminium/ 0.1-2.5 M NaOH/ Langmuir and Freundlich model | GA, GM, TM, ES, PM | Gum Arabic (GA) treated and tested as potential corrosion inhibitor on aluminum corrosion inhibition in 1 M NaOH by various parameters like hydrogen evolution, adsorption, kinetics, gravimetric etc. Thermometric analysis reveals the % inhibition of 75.86 at 0.5 g/L amount of inhibitor | 65 |

8	 gum arabic Guar gum	Aluminium/ 2 M HCl/ Temkin model	WL, TM	S. A. Umoren and coworkers (2008) reveals that dacroydesedulis exudates gum shows % inhibition of 42.0 at 5 g/L concentration (30°C temperature)	66
9	 carboxymethyl cellulose	Al/ Ground water (pH 11)/ Mixed (cathodic basically)	SEM, EIS, WL, PDP	Researchers reported the CMC as mixed type inhibitor as per WL analysis	67
10		Al & Al-Si alloys/ 0.1 M NaOH/ Anodic	PDP, CV, EIS	Analysis like CV, EIS and WL applied for Al & Al-Si alloys corrosion at 0.1 M NaOH and find efficient towards corrosion inhibition	68

11	Pectic Acid	Al/ 4 M NaOH	HE, WL	Scientists proved PA as efficient inhibitor by applying HE and WL techniques	69
12	Pectic Acid	Al/ 4 M NaOH	HE, WL	WL measurement suggest the good inhibition property of PA which follows Langmuir model	70
13	Chitosan	Cu/ 0.5 M HCl/ Mixed type	PDP, DFT, WL, EFM, EIS	Chitosan studied by applying PDP, DFT, WL, EFM, EIS techniques as Cu corrosion inhibitor and reported as mixed type nature	71
14	Chitosan	304 steel/Cu sheets/ 2% acetic acid/ Mixed type	EIS, FTIR, TGA, SEM, PDP	Various techniques like EIS, FTIR, TGA, SEM, PDP applied and reported the mixed type behavior of chitosan	72
15	carrageenan alginate	H₂SO₄ and HCl/Mixed Type/ Langmuir model	GA, WL, EIS, SEM, PDP, OCP, DFT, XPES, XRD, FTIR, AFM	Various techniques applied and result reveals the mixed type inhibition through the polysaccharides	73

3. Adsorption mechanism

Organic inhibitor molecules have been used in place of organic compounds such as phosphate, chromate, and arsenate over the past decades to prevent metal dilution in aqueous solutions [74]. Polysaccharides and their derivatives are non-polluting organic compounds that are used as effective corrosion inhibitors [50-54]. These organic materials in the corrosive solution are adsorbed onto the metal surface by physical force

Sustainable Corrosion Inhibitors   Materials Research Forum LLC
Materials Research Foundations**107** (2021) 70-100   https://doi.org/10.21741/9781644901496-4

(physisorption) or chemical bond (chemical adsorption). These organic inhibitors reduce their anodic resolution rate by covering the entire metal surface or preventing cathodic reaction by blocking the active site of the metal surface so that the metal surface is not damaged or degraded in any way [75]. In this way, the polymer is deposited on to the metal surface like a blocker, in which corrosive molecule cannot attack. The efficiency of corrosion inhibition of polysaccharides depends on different types of inhibitor molecules. Some molecules have a greater number of cyclic rings, some have functional groups and alkyl chain lengths, and some have sulfonate and carboxylic groups [76]. In addition to the bulky group in the inhibitor molecule, the presence of unshared electron pair and extended π bond makes the organic molecule as the efficient corrosion inhibitor. Inhibitor molecules that have ΔG values less than - 20 kJ / mol^{-1} are physisorption and ΔG values up to -40kJ/ mol^{-1} are adjusted on the metal surface by chemical adsorption. The nature of the inhibitor representing the value of ΔG between these two is considered as mixed type [77-81]. Figs. 2 and 3 represents the proposed adsorption mechanism of chemisorption and physisorption, respectively.

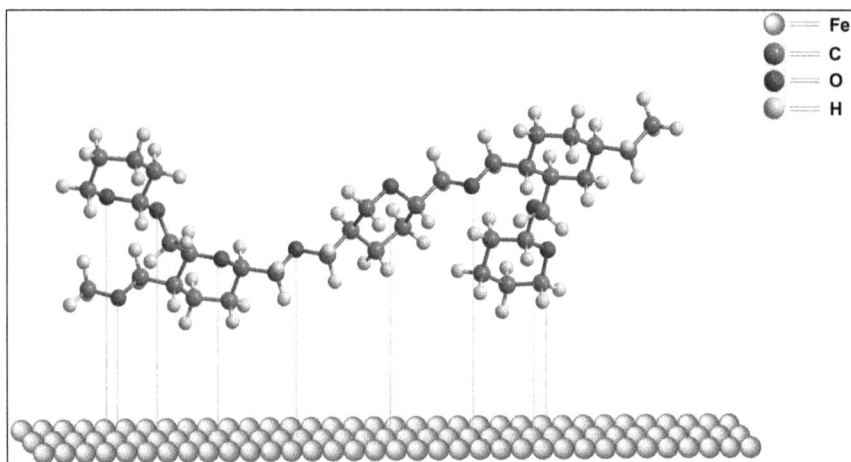

Figure 2.   Proposed chemisorption mechanism of polysaccharide.

Sustainable Corrosion Inhibitors   Materials Research Forum LLC
Materials Research Foundations**107** (2021) 70-100   https://doi.org/10.21741/9781644901496-4

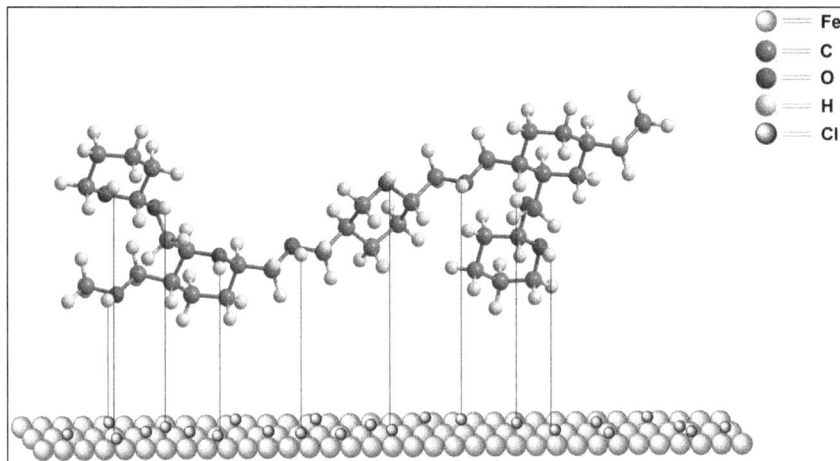

Figure 3.   Proposed physisorption mechanism of polysaccharide.

4.   Further aspects

Polysaccharides based corrosion inhibitor suggests that most of the studies have been on mild steel / steel and aluminum metal. Mainly HCl, H_2SO_4 and NaCl have been frequently used as electrolytic media. But there has been almost no research on metal like Cu, Zn, Mg, Sn, Pb and their metal alloys as a corrosion inhibitor of Polysaccharides, on which research is yet to be done. Theoretical studies like MD simulation and DFT calculation applied in limited research papers, which is yet to be done. Electrolytic medium has been used in low concentration. While the behaviors of corrosion inhibitor in high concentration has been studied in few papers. Also, other media other than NaCl can be used for basic medium. In this way, there is still research into polysaccharides corrosion inhibitors is to be done.

Conclusion

Polysaccharides are polymers containing repeated units of carbohydrate sugar. These complex structures are very good corrosion inhibitors due to biodegradability, easy availability, cost effective, non-toxic, environment friendly properties. The cyclic rings, odd atoms and π electron densities present in them increase the efficiency, allowing them on to deposit above the metal surface. Some natural green synthetic biopolymer, mixed type inhibitors including composites of polysaccharides are described in this book chapter. In addition to chemical and electrochemical analysis, computational studies such as DFT and MD simulations are also described in some recent research papers, which are included in this book chapter. Based on the properties of the analysis and the molecule, it would be fair to say that polysaccharide is a good alternative to toxic, expensive, and hazardous inhibitors.

Acknowledgement

Authors greatly acknowledge the Principal Govt. Digvijay Autonomous College, Rajnandgaon for providing basic facilities.

Abbreviations

WL	=	Weight loss
EIS	=	Electrochemical impedance spectroscopy
PDP	=	Potentiodynamic polarization
OCP	=	Open circuit potential
SEM	=	Scanning electron microscopy
EDS	=	Electron dispersion x-ray spectroscopy
QCC	=	Quantum chemical calculation
DFT	=	Density function theory,
XPES	=	X-ray photoelectron spectroscopy
MDS	=	Molecular dynamic simulation
XRD	=	X-ray diffraction
FTIR	=	Furrier transform infrared
AFM	=	Atomic force microscopy
EDXA	=	Energy dispersive x-ray analysis
GCMS	=	Gas chromatography mass spectrophotometer
HE	=	Hydrogen evolution
CTAB	=	Cetyl Trimethyl Ammonium Bromide

References

[1] C. Verma, M.A. Quraishi, Thermodynamic, electrochemical and surface studies of dendrimers as effective corrosion inhibitors for mild steel in 1 M HCl, Anal. Bioanal. Electrochem. 8(1) (2016) 104-123.

[2] S. Bourichi, Y.K. Rodi, M. El Azzouzi, Y. Kharbach, F.O. Chahdi, A. Aouniti, Inhibitive effect of new synthetized imidazopyridine derivatives for the mild steel corrosion in Hydrochloric acid medium, J. Mater. Environ. Sci. 8(5) (2017) 1696-1707.

[3] E.S. Ferreira, C. Giacomelli, F.C. Giacomelli, A. Spinelli, Evaluation of the inhibitor effect of L-ascorbic acid on the corrosion of mild steel, Mater. Chem. Phys. 83(1) (2004) 129-134. https://doi.org/10.1016/j.matchemphys.2003.09.020

[4] I. Sekine, Y. Nakahata, H. Tanabe, The corrosion inhibition of mild steel by ascorbic and folic acids, Corros. Sci. 28(10) (1988) 987-1001. https://doi.org/10.1016/0010-938X(88)90016-9

[5] L.R. Chauhan, G. Gunasekaran, Corrosion inhibition of mild steel by plant extract in dilute HCl medium, Corros. Sci. 49 (2007) 1143-1161. https://doi.org/10.1016/j.corsci.2006.08.012

[6] X. Wang, Y. Gu, Q. Zhang, L. Xu, X. Li, Rose, gardenia, and solanumviolaceum extracts as inhibitors of steel corrosion, Int. J. Electrochem. Sci., 14 (2019) 8405-8418. https://doi.org/10.20964/2019.09.41

[7] S. Paul, I. Koley, Corrosion inhibition of carbon steel in acidic environment by papaya seed as green inhibitor, J. Bio. Tribo. Corros. 2 (2016) 6. https://doi.org/10.1007/s40735-016-0035-2

[8] R. Anitha, S. Chitra, Corrosion resistance of cissusquadrangularis extracts on metal in aggressive medium: Gravimetric and surface examinations, Rasayan J. Chem. 12 (2019) 1326-1339. https://doi.org/10.31788/RJC.2019.1235195

[9] N. Raghavendra, J. IshwaraBhat, An environmentally friendly approach towards mitigation of Al corrosion in hydrochloric acid by yellow colour ripe arecanut husk extract: Introducing potential and sustainable inhibitor for material protection, J. Bio. Tribo. Corros. 4 (2018). https://doi.org/10.1007/s40735-017-0112-1

[10] A.K. Satapathy, G. Gunasekaran, S.C. Sahoo, K. Amit, P.V. Rodrigues, Corrosion inhibition by Justiciagendarussa plant extract in hydrochloric acid solution, Corros. Sci. 51 (2009) 2848-2856. https://doi.org/10.1016/j.corsci.2009.08.016

[11] H. Mohammed, S.bt. Sobri, Corrosion inhibition studies of cashew nut (Anacardiumoccidentale) on carbon steel in 1.0 M hydrochloric acid environment, Mater. Lett.. 229 (2018) 82-84. https://doi.org/10.1016/j.matlet.2018.06.108

[12] E.S. Ferreira, C. Giacomelli, F.C. Giacomelli, A. Spinelli, Evaluation of the inhibitor effect of L-ascorbic acid on the corrosion of mild steel, Mater. Chem. Phys. 83 (2004) 129-134. https://doi.org/10.1016/j.matchemphys.2003.09.020

[13] P.B. Raja, M.G. Sethuraman, Natural products as corrosion inhibitor for metals in corrosive media - A review, Mater. Lett. 62 (2008) 113-116. https://doi.org/10.1016/j.matlet.2007.04.079

[14] M. Behzadnasab, S.M. Mirabedini, K. Kabiri, S. Jamali, Corrosion performance of epoxy coatings containing silane treated ZrO2 nanoparticles on mild steel in 3.5% NaCl solution, Corros. Sci. 53 (2011) 89-98. https://doi.org/10.1016/j.corsci.2010.09.026

[15] X. Li, S. Deng, H. Fu, G. Mu, Inhibition effect of 6-benzylaminopurine on the corrosion of cold rolled steel in H2SO4 solution, Corros. Sci. 51 (2009) 620-634. https://doi.org/10.1016/j.corsci.2008.12.021

[16] W.H. Li, Q. He, S.T. Zhang, C.L. Pei, B.R. Hou, Some new triazole derivatives as inhibitors for mild steel corrosion in acidic medium, J. Appl. Electrochem. 38 (2008) 289-295. https://doi.org/10.1007/s10800-007-9437-7

[17] A. Popova, E. Sokolova, S. Raicheva, M. Christov, AC and DC study of the temperature effect on mild steel corrosion in acid media in the presence of benzimidazole derivatives, Corros. Sci. 45 (2003) 33-58. https://doi.org/10.1016/S0010-938X(02)00072-0

[18] A. Döner, R. Solmaz, M. Özcan, G. Kardaş, Experimental and theoretical studies of thiazoles as corrosion inhibitors for mild steel in sulphuric acid solution, Corros. Sci. 53 (2011) 2902-2913. https://doi.org/10.1016/j.corsci.2011.05.027

[19] R. Solmaz, G. Kardaş, M. Çulha, B. Yazici, M. Erbil, Investigation of adsorption and inhibitive effect of 2-mercaptothiazoline on corrosion of mild steel in hydrochloric acid media, Electrochim. Acta. 53 (2008) 5941-5952. https://doi.org/10.1016/j.electacta.2008.03.055

[20] H. Ashassi-Sorkhabi, B. Shaabani, D. Seifzadeh, Corrosion inhibition of mild steel by some schiff base compounds in hydrochloric acid, Appl. Surf. Sci. 239 (2005) 154-164. https://doi.org/10.1016/j.apsusc.2004.05.143

[21] M. Lebrini, M. Lagrenée, H. Vezin, L. Gengembre, F. Bentiss, Electrochemical and quantum chemical studies of new thiadiazole derivatives adsorption on mild steel in normal hydrochloric acid medium, Corros. Sci. (2005) 47, 485-505. https://doi.org/10.1016/j.corsci.2004.06.001

[22] M. Palomar-Pardavé, M. Romero-Romo, H. Herrera-Hernández, M.A. Abreu-Quijano, N.V. Likhanova, J. Uruchurtu, J.M. Juárez-García, Influence of the alkyl chain length of 2 amino 5 alkyl 1,3,4thiadiazole compounds on the corrosion inhibition of steel immersed in sulfuric acid solutions, Corros. Sci. 54 (2012) 231-243. https://doi.org/10.1016/j.corsci.2011.09.020

[23] A.S. Begum, J. Mallika, P. Gayathri, Corrosion inhibition property of some 1, 3, 4-thiadiazolines on mild steel in acidic medium, E. J. Chem. 7 (2010) 185-197. https://doi.org/10.1155/2010/623298

[24] I.O. Arukalam, Durability and synergistic effects of KI on the acidcorrosion inhibition of MS by hydroxypropyl methylcellulose, Carbohydr. Polymer. 112 (2014) 291-9. https://doi.org/10.1016/j.carbpol.2014.05.071

[25] M. Abdallah, Guar Gum as Corrosion Inhibitor for Carbon Steel in Sulfuric Acid Solutions, Port. ElectrochimicaActa (2003) 161-175. https://doi.org/10.4152/pea.200402161

[26] M. Mobin, M.A. Khan, M. Parveen, Inhibition of MS Corrosion in Acidic Medium Using Starch and Surfactants Additives, J. Appl. Polymer Sci. 121(3) (2011) 1558-1565. https://doi.org/10.1002/app.33714

[27] M. Mobin, M.A. Khan, Investigation on the Adsorption and Corrosion Inhibition Behavior of Gum Acacia and Synergistic Surfactants Additives on MS in 0.1 M H2SO4, J. Disper. Sci. Technol. 34(11) (2013) 1496-1506. https://doi.org/10.1080/01932691.2012.751031

[28] M. Mobin, M. Rizvi, Inhibitory effect of xanthan gum and synergistic surfactant additives for MS corrosion in 1 M HCl, Carbohydr. Polymer. 136 (2016) 384-393. https://doi.org/10.1016/j.carbpol.2015.09.027

[29] M. Mobin, M. Rizvi, Polysaccharide from Plantagoas a Green Corrosion Inhibitor for Carbon Steel in 1M HCl Solution, Carbohydr. Polymer. 160 (2017) 172-183. https://doi.org/10.1016/j.carbpol.2016.12.056

[30] M.V. Fiori-Bimbi, P.E. Alvarez, H. Vaca, C.A. Gervasi, Corrosion inhibition of MS in HCl solution by pectin, Corros. Sci. 92 (2015) 192-199. https://doi.org/10.1016/j.corsci.2014.12.002

[31] M.M. Solomon, S.A. Umoren, I.I. Udosoro, A.P. Udoh, Inhibitive and adsorption behaviour of carboxymethyl cellulose on MS corrosion in sulphuric acid solution, Corros. Sci. 52(4) (2010) 1317-1325. https://doi.org/10.1016/j.corsci.2009.11.041

[32] M.N. EL-Haddad, Hydroxyethylcellulose used as an eco-friendly inhibitor for 1018 c- steel corrosion in 3.5% NaCl solution, Carbohydr. Polymer. 112 (2014) 595-602. https://doi.org/10.1016/j.carbpol.2014.06.032

[33] M. Bello, N. Ochoa, V. Balsamo, F. López-Carrasquero, S. Coll, A. Monsalve, G. González, Modified cassava starches as corrosion inhibitors of carbon steel: An

electrochemical and morphological approach, Carbohydr. Polymer. 82(3) (2010) 561-568. https://doi.org/10.1016/j.carbpol.2010.05.019

[34] M. Mobin, M. Rizvi, L.O. Olasunkanmi, E.E. Ebenso, Biopolymer from tragacanth gum as a green corrosion inhibitor for carbon steel in 1 M HCl solution, ACS Omega. 2(7) (2018) 3997–4008. https://doi.org/10.1021/acsomega.7b00436

[35] M. Mobin, M. Rizvi, Adsorption and corrosion inhibition behavior of hydroxyethylcellulose and synergistic surfactants additives for carbon steel in 1 M HCl, Carbohydr. Polymer 156 (2017) 202-214. https://doi.org/10.1016/j.carbpol.2016.08.066

[36] N.O. Eddy, P. Ameh, C.E. Gimba , E.E. Ebenso, Chemical information from gcms of Ficus platyphylla gum and its corrosion inhibition potential for MS in 0.1 M HCl, Int. J. Electrochem. Sci. 7 (2012) 5677 - 5691.

[37] N.O. Eddy, P. Ameh, C.E. Gimba, E.E. Ebenso, GCMS studies on Anogessus Leocarpus (Al) gum and their corrosion inhibition potential for MS in 0.1 M HCl, Int. J. Electrochem. Sci. 6 (2011) 5815 - 5829.

[38] N.O. Eddy, A.O. Odiongenyi, P.O. Ameh, E.E. Ebenso, Corrosion inhibition potential of Daniella Oliverri Gum exudate for MS in acidic medium, Int. J. Electrochem. Sci., 7 (2012) 7425 - 7439.

[39] P.O. Ameh, Adsorption and Inhibitive Properties of Khaya ivorensis Gum for the Corrosion of MS in HCl, Int. J. Mod. Chem. 2(1) (2012) 28-40. https://doi.org/10.1007/s11164-013-1117-0

[40] S. Banerjee, V. Srivastava, M.M. Singh, Chemically modified natural polysaccharide as green corrosion inhibitor for MS in acidic medium, Corros. Sci. 59 (2012) 35-41. https://doi.org/10.1016/j.corsci.2012.02.009

[41] V. Rajeswari, D. Kesavan, M. Gopiraman, P. Viswanathamurthi, Physicochemical studies of glucose, gellan gum, and hydroxypropyl cellulose-Inhibition of cast iron corrosion, Carbohydr. Polymer. 95(1) (2013) 288- 294. https://doi.org/10.1016/j.carbpol.2013.02.069

[42] M. Abdallah, Guar gum as corrosion inhibitor for carbon steel in sulfuric acid solutions, Port. ElectrochimicaActa 22 (2004) 161-175. https://doi.org/10.4152/pea.200402161

[43] H. Bentrah, Y. Rahali, A. Chala, Gum Arabic as an eco-friendly inhibitor for API 5L X42 pipeline steel in HCl medium, Corros. Sci. 82 (2014) 426-431. https://doi.org/10.1016/j.corsci.2013.12.018

[44] A. Biswas, S. Pal, G. Udayabhanu, Experimental and theoretical studies of xanthan gum and its graftco-polymer as corrosion inhibitor for MS in 15% HCl, Appl. Surf. Sci. 353 (2015) 173-183. https://doi.org/10.1016/j.apsusc.2015.06.128

[45] M.M. Solomon, S.A. Umoren, I.I. Udosoro, A.P. Udoh, Inhibitive and adsorption behaviour of carboxymethyl cellulose on MS corrosion in sulphuric acid solution, Corros. Sci. 52(4) (2010) 1317-1325. https://doi.org/10.1016/j.corsci.2009.11.041

[46] S.A. Umoren, U.F. Ekanem, Inhibition of MS corrosion in H2SO4 using exudate gum from pachylobusedulis and synergistic potassium halide additives, Chem. Eng. Comm. 197(10) (2010) 1339-1356. https://doi.org/10.1080/00986441003626086

[47] N. Manimaran, S. Rajendran, M. Manivannan, J.A. Thangakani, A.S. Prabha, Corrosion inhibition by carboxymethyl cellulose, Eur. Chem. Bull. 2(7) (2013) 494-498.

[48] M.N. EL-Haddad, Hydroxyethylcellulose used as an eco-friendly inhibitor for 1018 c- steel corrosion in 3.5% NaCl solution, Carbohydr. Polymer. 112 (2014) 595-602. https://doi.org/10.1016/j.carbpol.2014.06.032

[49] I.O. Arukalam, I.C. Madufor, O. Ogbobe, E.E. Oguzie, Inhibition of mild steel corrosion in sulphuric acid medium by hydroxyethyl cellulose, Chem. Eng. Commun. 202(1) (2014) 112-122. https://doi.org/10.1080/00986445.2013.838158

[50] T. Brindha, J. Mallika, V.S. Moorthy, Synergistic effect between starch and substituted piperidin-4-one on the corrosion inhibition of MS in acidic medium, J. Mater. Environ. Sci. 6(1) (2015) 191-200.

[51] M. Prabakaran, S. Ramesh , V. Periasamy, B. Sreedhar, The corrosion inhibition performance of pectin with propyl phosphonic acid and Zn2+ for corrosion control of carbon steel in aqueous solution, Res. Chem. Intermed. 41 (2015) https://doi.org/10.1007/s11164-014-1558-0

4649-4671.

[52] R. Geethanjali, A.A.F. Sabirneeza, S. Subhashini, Water-soluble and biodegradable pectin-grafted polyacrylamide and pectin-grafted polyacrylic acid: electrochemical investigation of corrosion-inhibition behaviour on MS in 3.5% NaCl media, Indian J. Mater. Sci. (2014) 356075. https://doi.org/10.1155/2014/356075

[53] S.A. Umoren, M.J. Banera, T. Alonso-Garcia, C.A. Gervasi, M.V. Mırı'fico, Inhibition of MS corrosion in HCl solution using chitosan, Cellulose. 20 (2013) 2529-2545. https://doi.org/10.1007/s10570-013-0021-5

[54] A.M. Fekry, R.R. Mohamed, Acetyl thiourea chitosan as an eco-friendly inhibitor for MS in sulphuric acid medium, Electrochim. Acta. 55(6) (2010) 1933-1939. https://doi.org/10.1016/j.electacta.2009.11.011

[55] A.M. Atta, G.A. El-Mahdy, H.A. Al-Lohedan, A.R.O. Ezzat, Synthesis of nonionic amphiphilic chitosan nanoparticles for active corrosion protection of steel, J. Mol. Liq. 211 (2015) 315-323. https://doi.org/10.1016/j.molliq.2015.07.035

[56] S. John, A. Joseph, A.J. Jose, B. Narayana, Enhancement of corrosion protection of MS by chitosan/ZnO nanoparticle composite membranes, Prog. Org. Coatings. 84 (2015) 28-34. https://doi.org/10.1016/j.porgcoat.2015.02.005

[57] S.H.D. Koesoemo, R. Ruriyanti, L.S. Anggara, Application chitosan derivatives as inhibitor corrosion on steel with fluidization method, J. Chem. Pharm. Res. 7 (2015) 260-267.

[58] Y. Sangeetha, S. Meenakshi, C.S. Sundaram, Corrosion mitigation of N-(2-hydroxy-3-trimethyl ammonium)propylchitosan chloride as inhibitor on MS, Int. J. Biol. Macromol. 72 (2015) 1244- 1249. https://doi.org/10.1016/j.ijbiomac.2014.10.044

[59] I.O. Arukalam, N.T. Ijomah, S.C. Nwanonenyi, H.C. Obasi, B.C. Aharanwa, P.I. Anyanwu, Studies on acid corrosion of aluminium by a naturally occurring polymer (Xanthan gum), Int. J. Sci. Eng. Res. 5(3) (2014) 663-673.

[60] N.O. Eddy, P.O. Ameh, M.Y. Gwarzo, I.J. Okop, S.N. Dodo, Physicochemical study and corrosion inhibition potential of ficus tricopoda for aluminium in acidic medium, Port. Electrochim. Acta. 31(2) (2013) 79-93. https://doi.org/10.4152/pea.201302079

[61] N.O. Eddy, P.O. Ameh, A.O. Odiongenyi, Physicochemical characterization and corrosion inhibition potential of ficus benjamina (fb) gum for aluminum in 0.1 M H2SO4, Port. Electrochim. Acta. 32(3) (2014) 183-197. https://doi.org/10.4152/pea.201403183

[62] N.O. Eddy, U.J. Ibok, P.O. Ameh, N.O. Alobi, M.M. Sambo, Adsorption and quantum chemical studies on the inhibition of the corrosion of aluminum in HCl by Gloriosasuperba (GS) Gum, Chem. Eng. Comm. 201(10) (2014) 1360-1383. https://doi.org/10.1080/00986445.2013.809000

[63] N.O. Eddy, P.O. Ameh, O. Danclementino, A. Odiongenyi, Adsorption and chemical studies on the inhibition of the corrosion of aluminium in hydrochloric acid by commiphora africana gum, Int. J. Chem. Mater. Environ. Res. 1(1) (2014) 16-28.

[64] R. Rosliza, W.B.W. Nik, Improvement of corrosion resistance of AA6061 alloy by tapioca starch in seawater, Curr. Appl. Phys. 10(1) (2010) 221-229. https://doi.org/10.1016/j.cap.2009.05.027

[65] S.A. Umoren, I.B. Obot, E.E. Ebenso, P.C. Okafor, O. Ogbobe, E.E. Oguzie, Gum arabic as a potential corrosion inhibitor for aluminium in alkaline medium and its adsorption characteristics, Anti-Corros. Meth. Mater. 53(5) (2006) 277-282. https://doi.org/10.1108/00035590610692554

[66] S.A. Umoren, I.B. Obot, E.E. Ebenso, N. Obi-Egbedi, Studies on the inhibitive effect of exudate gum from dacroydesedulison the acid corrosion of aluminium, Port. Electrochim. Acta (2008) 199-209. https://doi.org/10.4152/pea.200802199

[67] R. Kalaivani, P.T. Arasu, S. Rajendran, Inhibitive nature of Carboxymethylcellulose with Zn2+ ion, Chem. Sci. Trans. 2(4) (2013) 1352-1357. https://doi.org/10.7598/cst2013.618

[68] S. Eid, M. Abdallah, E.M. Kamar, A.Y. El-Etre, Corrosion inhibition of aluminum and aluminum silicon alloys in sodium hydroxide solutions by methyl cellulose, J. Mater. Environ. Sci. 6(3) (2015) 892-901.

[69] I. Zaafarany, Corrosion inhibition of aluminum in aqueous alkaline solutions by alginate and pectate water-soluble natural polymer anionic polyelectrolytes, Port. Electrochim. Acta. 30(6) (2012) 419-426. https://doi.org/10.4152/pea.201206419

[70] R.M. Hassan, I.A. Zaafarany, Kinetics of corrosion inhibition of aluminum in acidic media by water-soluble natural polymeric pectates as anionic polyelectrolyte inhibitors, Mater. (Basel) 6(6) (2013) 2436-2451. https://doi.org/10.3390/ma6062436

[71] M.N. El-Haddad, Chitosan as a green inhibitor for copper corrosion in acidic medium, Int. J. Biol. Macromol. 55 (2013) 142- 149. https://doi.org/10.1016/j.ijbiomac.2012.12.044

[72] M.L Li, R.H. Li, J. Xu, X. Han, T.Y. Yao, J. Wang, Thiocarbohydrazide-modified chitosan as anticorrosion and metal ion adsorbent, J. Appl. Polym. Sci. 131(17) (2014) 40671. https://doi.org/10.1002/app.40671

[73] S.A. Umoren1, U.M. Eduok, Application of carbohydrate polymers as corrosion inhibitors for metal substrates in different media: A review, Carbohydr. Polymer. 140 (2016) 314-341. https://doi.org/10.1016/j.carbpol.2015.12.038

[74] J. Aljourani, K. Raeissi, M.A. Golozar, Benzimidazole and its derivatives as corrosion inhibitors for mild steel in 1M HCl solution, Corros. Sci. 51 (2009) 1836-1843. https://doi.org/10.1016/j.corsci.2009.05.011

Materials Research Forum LLC
https://doi.org/10.21741/9781644901496-4

[75] A. Popova, E. Sokolova, S. Raicheva, M. Christov, AC and DC study of the temperature effect on mild steel corrosion in acid media in the presence of benzimidazole derivatives, Corros. Sci. 45 (2003) 33-58. https://doi.org/10.1016/S0010-938X(02)00072-0

[76] H. Cang, Z. Fei, H. Xiao, J. Huang, Q. Xu, Inhibition effect of reed leaves extract on steel in hydrochloric acid and sulphuric acid solutions, Int. J. Electrochem. Sci. 7 (2012) 8869-8882.

[77] C. Kamal, M.G. Sethuraman, Spirulinaplatensis - A novel green inhibitor for acid corrosion of mild steel, Arab. J. Chem. 5 (2012) 155-161. https://doi.org/10.1016/j.arabjc.2010.08.006

[78] M. Bouklah, B. Hammouti, M. Lagrenée, F. Bentiss, Thermodynamic properties of 2,5-bis(4-methoxyphenyl)-1,3,4-oxadiazole as a corrosion inhibitor for mild steel in normal sulfuric acid medium, Corros. Sci. 48 (2006) 2831-2842. https://doi.org/10.1016/j.corsci.2005.08.019

[79] F.B. Mainier, R.de.M.B.e. Silva, Evaluation of corrosion inhibitors in acid medium, Anti Corros. Meth. Mater. 62 (2015) 241-245. https://doi.org/10.1108/ACMM-12-2013-1329

[80] M. Hosseini, S.F.L. Mertens, M. Ghorbani, M.R. Arshadi, Asymmetrical Schiff bases as inhibitors of mild steel corrosion in sulphuric acid media, Mater. Chem. Phys. 78 (2003) 800-808. https://doi.org/10.1016/S0254-0584(02)00390-5

[81] J. Zhou, X. Niu, Y. Cui, Z. Wang, J. Wang, R. Wang, Study on the film forming mechanism, corrosion inhibition effect and synergistic action of two different inhibitors on copper surface chemical mechanical polishing for GLSI, Appl. Surf. Sci. 505 (2019) 144507. https://doi.org/10.1016/j.apsusc.2019.144507

Sustainable Corrosion Inhibitors   Materials Research Forum LLC
Materials Research Foundations**107** (2021) 101-129   https://doi.org/10.21741/9781644901496-5

Chapter 5

Green Organic Inhibitors for Corrosion Protection

Y.V.D. Nageswar[1]*, V.J. Rao[2]

[1]Retired chief scientist, CSIR - Indian Institute of Chemical Technology, Hyderabad - 500 007, India

[2]Emeritus Scientist & Advisory Consultant, Hetero Research Foundation, Balanagar, Hyderabad – 500018, India

* dryvdnageswar@gmail.com

Abstract

Plants are a rich source of different varied organic compounds. Due to the important applications of naturally occurring chemicals their derivatives are also pursued for modifying and potentiating the activities of natural products. Metallic corrosion is a natural process resulting in heavy losses in various fields. Non hazardous and non toxic corrosion inhibitors gained significance due to the environmental regularities and guidelines issued in the course of saving the pristine nature of environment and to maintain the sustainability of our earth. Green corrosion inhibitors play a potential role for the above said cause. Recent research contributions on green corrosion inhibitors from the active researchers in the concerned expertise are presented briefly here to give an idea about the current research activity across the world.

Keywords

Phytochemicals, Surfactants, Protective Film, Inhibition Efficacy, Corrosion Inhibition, Physisorption, Chemisorption, Corrosion Rate, Inhibitor Concentration, Electrochemical Impedance Spectroscopy

Contents

1.   Introduction

Metals and alloys react with the surrounding environment and undergo corrosion overtime, causing loss of material and their property. Corrosion process caused by either cathodic or anodic process can be controlled by the application of corrosion inhibitor on the metal or alloy surface. Inhibitors are adsorbed on the metal surface to provide an active barrier layer between metal and degrading and aggressive environment. Corrosion inhibition can be initiated by coating the concerned metallic surface with organic or inorganic compounds which can stop the oxidation-reduction process. Many inhibitors are toxic and hazardous. These can be cathodic, anodic or mixed type.

Due to the current environment regularities and guidelines with an aim to maintain the sustainability of the environment and to reduce the toxicity of chemical applications presently research is being focussed to find greener and more effective corrosion inhibitors which can deliver higher corrosion inhibition efficiency. Many plant extracts, compounds derived from natural products, synthetic organic compounds and ionic liquids come under this category. Plant extracts are generally rich in organic compounds containing functional groups, π-electrons in conjugation with double or triple bonds and the presence of hetero atoms like oxygen, nitrogen, or sulphur. During the adsorption process the active compounds present in the inhibitor layer interact with the metal surface involving electrons. Herbs like hibiscus, and thyme, saccharides like fructose, natural products like guar gum, nicotine, caffeine, eugenol, berberine, chamomile and gum Arabic were examined for their efficacy as corrosion inhibitors. Among metals steel, aluminium and copper were under study. Generally, the green corrosion inhibitors reduce or affect corrosion rate by increasing electrical resistance of the metal surface with the formation of a protective layer, by changing rate of anodic/cathodic reactions or by changing the diffusion rate of aggressive ions interacting with metallic structures.

The present review is intended to showcase the recent research findings on green corrosion inhibitors reported by enthusiastic researchers actively pursuing the field of corrosion inhibitors. The focus of the write-up is to expose the work from the angle of organic molecular structural point of view. Moreover, the available literature is classified into two parts, a. plant/animal extract applications, b. application of organic compounds for the convenience of readers.

2.   Examples from recent literature

2.1   Plant/animal extract applications

Rodriguez – Torres et al. examined [1] the extract of Prunus persica leaves for corrosion inhibition against AISI 1018 carbon steel in presence of 0.5 M sulphuric acid solution at 25 °C. It was reported that inhibitory efficiency was directly proportional to the inhibitor concentration, reaching the maximum effect at 600 ppm (97%). Authors observed negligible damage to the steel surface on protecting with P.Persica extract. P.Persica extract is a well-known Chinese plant, abundant in phenolic compounds such as hydroxycinnamic acid, anthocyanins, flavonoids and procyanidins. By GC-MS analysis authors identified several natural products in the extracts such as α-tocopherol, β -pinene, β -sitosterol, α-terpinine, squalene, β-myrcene, β-Cymene, α-Pinene and Linoleic acid. In view of the abundance of α -tocopherol and β -sitosterol as well as the experimental data obtained, these phytochemicals are considered to be responsible for the CIE against AISI 1018 carbon steel in H2SO4. To ensure that these phytochemical green corrosion inhibitors to be also eco-friendly, authors assessed the toxicity of these extracts.

G. Subramanian et al. examined [2] the efficacy of a new corrosion inhibitor – Pithecollobium dulce – leaves extract for mild steel surfaces in presence of 1 M HCl and 3.5% NaCl solutions with the help of electrochemical techniques and weight loss measurements. It was observed that the corrosion inhibition efficiency increased with increase in concentration of the inhibitor. Authors also concluded that the formation of Fe – inhibitor complex may be responsible for corrosion inhibition.

LeticiaA.L.Guedes et al. assessed [3] the corrosion inhibition efficiency of the tannins from bark of Acasia mearnsii for AA7075 – T_6 aluminium alloy in presence of 0.1 M HCl solution employing electrochemical impedance spectroscopy, scanning electronmicroscopy and potentiodynamic polarization techniques and observed that inhibition efficacy increased with increase in inhibitor concentration. It was concluded that by forming a layer on the aluminium surface, the tannate compound reduced the metal dissolution reaction.

Hicham Taoui and co-authors investigated [4] the inhibition effect of bark resin of Schinusmolle on the corrosion of API 5Lx70 pipeline steel in HCl solution using electrochemical impedance spectroscopy and potentiodynamic polarization techniques. It was reported that the resin exhibited about94% efficiency at $2gL^{-1}$. Probable mechanism was also proposed by the authors. The anti corrosive property of the bark resin was explained as due to the adsorption of bark resin constituents such as pinicolic acid, isomasticadenoic acid and isomasticadienonalic acid (Fig. 1) on the surface and also due to two types of interactions, such as physisorption and chemisorption. Authors concluded that the bark resin acted as mixed type inhibitor exhibiting geometric blocking effect, and the corrosion was controlled by charge transfer process.

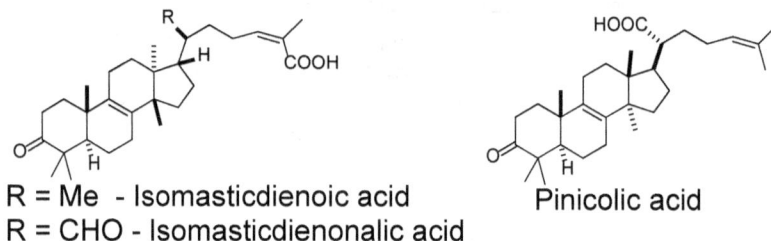

R = Me - Isomasticdienoic acid
R = CHO - Isomasticdienonalic acid

Pinicolic acid

Figure 1.   Structure of pinicolic acid, isomasticadenoic acid and isomasticadienonalic acid.

Aprael S.Yaro et al. described [5] apricot juice as a green corrosion inhibitor for mild steel surfaces in presence of 1M phosphoric acid solution at different temperatures. The studies proved that the corrosion rate was influenced by temperature and inhibitor concentration. Adsorption studies indicated that inhibitor was adsorbed on the metal surface. Authors also proved that apricot juice exhibited maximum inhibition efficiency of 75% at 30 °C.

Ahmed A. Farag and co-workers reported [6] that recovery shrimp waste protein acted as a green corrosion inhibitor for carbon steel of in presence of 1M HCl solution utilizing different electrochemical techniques. Authors also showed that the corrosion rate decreases with increase in the inhibitor concentration. Authors stated that the corrosion inhibition is due to the presence of donor atoms in the amino acids included in the protein structure.

Mohammad Tariq Saeed and co-workers demonstrated [7] that corrosion of mild steel can be inhibited by sweet melon peel extract in the presence of 1 M HCl solution with the help of potentiodynamic polarization and weight loss methods. Experiments were conducted with various extract concentrations at different temperatures. With 0.5 g/L concentration of the extract five times lower rate of corrosion was observed at 333 K.

Corrosion rate was directly related to the increase in temperatures from 295 to 333 K even in presence of the extract.

Anees A. Khadan and co-workers investigated [8] the use of Xanthiumstrumarium leaves extract for assessing the corrosion inhibition behaviour for low carbon steel in the presence of 1 M HCl solution and found that 94.82% inhibition efficiency at higher level of concentration and temperature. Authors observed that the efficiency increases with increase in temperature and concentration. FTIR studies indicated that the extract contained mixture of compounds such as amides, amines, aromatic and organic acids.

Narasimha Raghavendra and Jathi Iswara Bhat reported [9] red areca nut seed extract containing electron rich polyphenols, flavonoids, alkaloids and tannins as a green sustainable corrosion inhibitor for aluminium in presence of 0.5 M HCl solution utilizing AFM, SEM, electrochemical and weight loss studies. It was observed that protection efficiency is directly proportional to the concentration of the extract and inversely proportional to the solution temperature, and aluminium contact time in the test solution. Author described that the extract reduced the speed of aluminium corrosion by charge transfer process by the adsorption of plant constituents on the metal surface.

Ivan Pradipta et al. investigated [10] corrosion inhibition efficiency of green tea extract, a generous source of natural anti oxidants against steel reinforcing bars (rebars) embedded in mortar and also conducted a comparative study with commercial inorganic calcium nitrite corrosion inhibition at similar concentrations and equal volumes. It was observed that at similar volumes green tea exhibited higher efficiency than inorganic counterpart, attributed to marked increase in rebar polarisation resistance and reduction in iron oxidation rate. With the help of liquid chromatography – tandem mass spectrometry and LCMS studies, authors hypothesized that catechin or (-) epicatechin, (-) epicatechingallate and (-) epigallocatechin gallate are the source for the anti corrosion property of green tea extract as mixed type.

Authors Idouhli et al. evaluated [11] the corrosion inhibitory effect of Senecioanteu-phorbium extract containing macrocyclic diesters on S300 steel in presence of 1 M HCl solution utilizing electrochemical impedance spectroscopy and potentio-dynamic polarisation studies. SEM, energy dispersive X-ray spectroscopy (EDS) and contact angle analysis indicated the adsorption of organic compounds present in the extract on metal surface and the formation of protective film. Authors noted that as the extract concentration increased, there was a significant improvement in the corrosion inhibition potency, reaching a maximum of 91% at 30mg/L.

Rodriguez et al. described [12] Crataegusmexicana as a green eco-friendly mixed type corrosion inhibitor for AISI 1018 carbon steel in presence of 0.5 M sulphuric acid.

Authors evaluated inhibitory efficiency of methanolic extract of the leaves of Cmexicana at different concentrations from 200 to 500 ppm, with the help of potentio-dynamic polarization curves (PPC), electrochemical impedance spectroscopy (EIS) and weight loss methods. From the results of FTIR spectroscopy the presence of flavonoids in the plant extract was proved.

Saviour Umoren et al. communicated [13] experimental studies related to the corrosion inhibition efficiency of date palm seed extract for mild steel in presence of 1 M HCl as well as 0.5 M sulphuric acid solutions using electrochemical and weight loss methods. Authors proved that the corrosion efficiency increased with increase in extract concentration and decreased with increase in temperature. It was observed that the extract exhibited better corrosion inhibition for steel in presence of HCl solution than in sulphuric acid solution. Maximum effect was noticed in 1 M HCl and 0.5 M sulphuric acid solutions, respectively. Immersion time had also considerably influenced the efficiency. Adsorption of extract components on the metal surface afforded the corrosion inhibition as proposed from the trend of results observed.

Pandian Bothi Raja et al. studied [14] the corrosion inhibition effects of alkaloid extracts of leaves and bark of Ochrosiaoppostifolia and isoreserpiline (Fig. 2), the major alkaloid isolated from the extracts, against mild steel in presence of 1 M HCl solution. The property was evaluated using electro chemical techniques, such as electrochemical impedance spectroscopy (EIS), potentiodynamic polarization (PDP) measurements and scanning electron microscopy (SEM). Isoreserpiline present in the extracts was found to be responsible for the inhibitory effect, as judged from the FT-IR and molecular modelling studies conducted by the authors. The data collected from FT-IR analysis indicated that indole moiety of isoreserpiline was involved in the co-ordination with the MS surface.

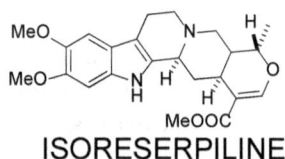

ISORESERPILINE

Figure 2.   Structure of Isoreserpiline.

Nur Izzah Nabilah Haris et al. evaluated [15] oil palm empty fruit bunch extract as green corrosion inhibitor for mild steel in 1 M HCl using weight loss technique. FT-IR and SEM studies indicated the presence of active chemical compounds in the deposited extracts on metal surface. Mechanism of inhibition was thought to incline more towards physical adsorption. It was considered that the presence of oxygen and nitrogen hetero atoms present in the chemical structures of the active molecules might be contributing towards inhibition apart from the presence of OH, C=C functional groups present in the molecules. Results of the study indicated great potential for the OPETB extract in the corrosion prevention for mild steel in presence of acidic environment.

Rajesh Haldhar et al. [16] investigated Valeriana Willichi roots extract as a green and sustainable corrosion inhibitor for mild steel in presence of 0.5 M sulphuric acid solution employing weight loss, electrochemical impedance spectroscopy, potentiodynamic polarization, SEM, AFM, FTIR quantum chemical calculations. Naphthoic acid, Iridoid and analogue are the important components in the extract and facilitate the adsorption on the metal surface. Authors reported maximum inhibition effecting up to 93.47% for mild steel at 500 mgL^{-1} concentration at 298 K in presence of 0.5 M sulphuric acid solution. Potentiodynamic polarization estimations proved the natural product of mixed type and quantum chemical calculations informed that naphthoic acid was the best corrosion inhibitor.

2.2   Application of organic compounds

Kegui Zhang and coworkers [17] synthesized two novel Schiff bases from renewable sources such as vanilline and amino acids – L-Lysine and L-Arginine by a green method (Fig. 3). Authors investigated the corrosion inhibitory efficiency of these compounds for mild steel immersed in 0.5 ml of 10^{-1}HCl solution and observed that both greatly reduced the corrosion rates. During the study authors utilized various experimental methods such as atomic force microscopy, X-ray photon electron microscopy, potentiodynamic polarization curves and electrochemical impedance spectroscopy.

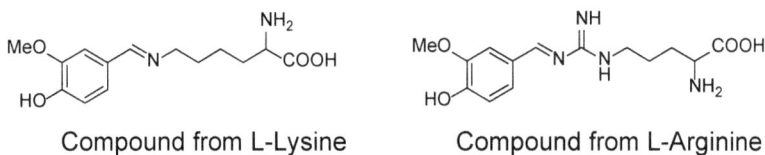

Compound from L-Lysine   Compound from L-Arginine

Figure 3.   Structure of compounds of L-Lysine and L-Arginine.

Olga Kaczerewska et al. investigated [18] six dimeric quaternary ammonium salts (gemini surfactants: Fig. 4) for corrosion inhibition property against stainless steel in 3M HCl solution with electrochemical impedance spectroscopy and potentiodynamic

Materials Research Forum LLC

https://doi.org/10.21741/9781644901496-5

polarization techniques. Authors observed best performance results with almost >95% of IE for surfactants with 12 carbon chain in the hydrophobic tail. Results indicated that adsorption process involved both physisorption and chemisorption.

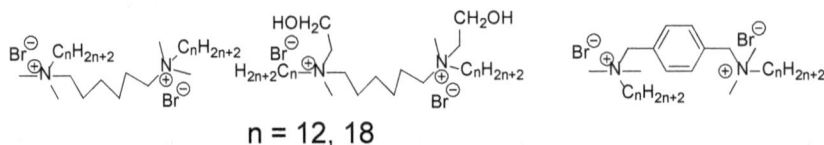

n = 12, 18

Figure 4.   Structure of dimeric quaternary ammonium salts.

Varvara et al. examined [19] four thiadiazole derivatives (Fig. 5) for their anticorrosive performances against naked and artificially patinated bronze in presence of acidic medium at pH3. The structures of thiadiazole derivatives were investigated by DFT using hybrid B3LYP, 6-31G models and possible inhibition mechanism of the protonated species on the bronze corrosion was proposed. Among the four compounds MMeT was proved to be most promising compound. In addition, in case of MMeT and MAT, it was observed that effectiveness increased with immersion time in acidic solution at pH 3.

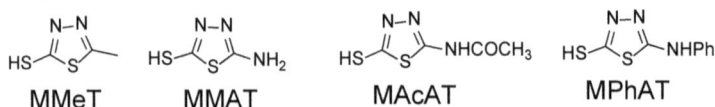

MMeT   MMAT   MAcAT   MPhAT

Figure 5.   Structure of thiadiazole derivatives.

Taha M.A. Eldebss and co-workers [20] synthesized a series of benzimidazole based heterocyclic derivatives embedded with a variety of functional groups (Fig. 6) and studied their copper corrosion inhibition property from elemental sulphur with in non-additive transformer oil following ASTM specified procedures. These compounds were prepared from N-methy-2-bromoacetyl benzimidazoles.

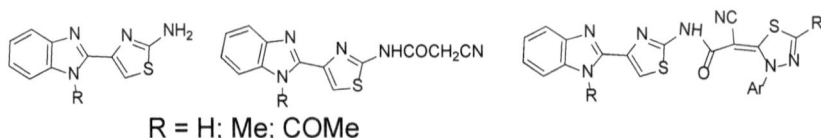

R = H; Me; COMe

Figure 6.   Structure of benzimidazole based heterocyclic derivatives.

Weiwei Zhang and co-authors [21] explained the synergistic inhibition effects of N-(furan-2ylmethyl)-7-H-purin-6-amine (Fig. 7) and iodide ion on mild steel corrosion in 1 mol/L HCl solution utilizing different techniques like surface analysis, weight loss tests, electrochemical measurements, potentiodynamic polarization experiments and electrochemical impedance spectroscopy. Authors observed that both FYPA and FYPA+KI acted as mixed type and exhibited temperature dependency. Plausible mechanism was proposed. It was also explained that FeI-FYPA+ protective film inhibited anodic dissolution of iron. These studies indicated that as the concentration increased, the corrosion inhibition efficiency increased and further increased in presence of KI.

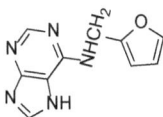

Figure 7.   Structure of N-(furan-2ylmethyl)-7-H-purin-6-amine.

Eva A.Yaqo et al. demonstrated [22] that an amino triazole derivative 3,5-di(4-hydroxyphenyl)-4-amino-1,2,4-triazole (Fig. 8) exhibited anti corrosive property for carbon steel in presence of kerosene, brine medium. The experiments revealed that it was of a mixed type inhibitor. Authors also examined the antibacterial action of the triazole compound against some types of corrosive bacteria. This amino triazole compound was reportedly achieved maximum of 86.74% efficiency at 100ppm concentration and exhibited temperature dependency. It was observed that efficiency increased with increase in concentration. Thus, the compound showed both anti-corrosive as well as antibacterial activity. Authors also discussed inhibition mechanism.

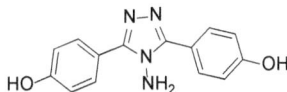

Figure 8.   Structure of an amino triazole derivative 3,5-di(4-hydroxyphenyl)-4-amino-1,2,4-triazole.

Hong Qin Liu and co-workers [23] synthesized some new alkyl hydroxyethylimidazolin salts from 2-alkylhydroxyethyl imidazoline and dimethylcarbonate, initially forming 2-alkylhydroxyethyl imidazolinemethylcarbonate (Fig. 9), which latter reacted with different acids.

$$R = C_{11}H_{23}; C_{13}H_{27}; C_{15}H_{31}$$
$$A = HCOO^{\ominus}; CH_3COO^{\ominus};$$
$$CH_3CH(OH)COO^{\ominus}$$

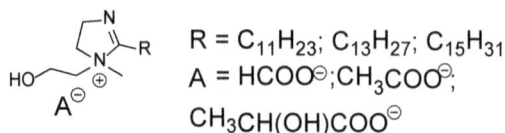

Figure 9.   Structure of 2-alkylhydroxyethyl imidazolinemethylcarbonate.

Authors observed that the compounds interacted with metals to form a protective layer on the metal surface to block the active sites decreasing the corrosion rates, and that the corrosion efficiency increased with increase in the length of the carbon chain. These exhibited anti corrosive property against mild steel in acid solutions.

Tuan K.A. Hoang et al. [24] developed sustainable gel electrolyte with pyrazole as an additive, which acted as a corrosion inhibitor and dendrite suppressor for aqueous Zn/LiMn$_2$O$_4$ battery.

Authors noted that battery in such condition exhibited high cyclability up to 85% capacity retention after 500 charge – discharge cycles at 4°C. It was reported that electrolyte containing 0.2 wt.% pyrazole exhibited superior results with 23% lower corrosion offering lowest absolute value of deposition.

Aiad and co-workers [25] prepared three cationic surfactants based on alginic acid (Fig. 10) and evaluated for corrosion inhibition for mild steel surfaces in presence of 1.0 M solution using different techniques such as electrochemical impedance spectroscopy, polarization and weight loss experiments. Authors observed that the efficiency was directly proportional to the hydrophobic chain length present in the synthesized compounds as well as the concentrations of the alginic cationic surfactants. The Polarization curves suggested that these behaved as mixed type inhibitors.

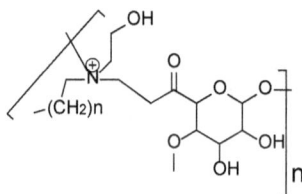

Figure 10.   Structure of cationic surfactants based on alginic acid.

An eco-friendly, one step, multi-component, ultrasound mediated synthesis of three pyrazolo-pyridine derivatives was described by Parul Dohare et al. [26] from hydrazine hydrate, ammonium acetate, β-dicarbonyl compound and appropriate aldehydes in presence of ethanol. Corrosion inhibition efficiency of these compounds was examined for mild steel in 1 M HCl solution with the help of electrochemical impedance spectroscopy, gravimetric and potentiodynamic polarization measurements. The adsorption behaviour and reactivity of these compounds was predicted using molecular dynamics and density functional theory. It was reported that these exhibited better performance (97% IE at 100 mgL^{-1}) than hexamine (90% IE at 100 mgL^{-1}). Among three compounds the one having methoxy substituent exhibited highest efficacy of 97.58% at 100 mgL^{-1} and was ascribed to the parallel adsorption on metal surface. These follow the ranking PP1 > PP2 > PP3 with mixed type inhibition.

Chandrabhan Verma et al. in their interesting communication [27] discussed the adsorption behaviour of glucosamine derived pyrimidine fused heterocyclic compounds (Fig. 11) as environmentally benign corrosion inhibitors for mild steel in presence of 1 M HCl solution. Authors utilized electrochemical, gravimetric, SEM, AFM, EDX

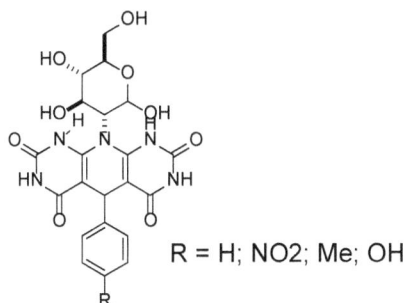

R = H; NO2; Me; OH

Figure 11.   Structure of glucosamine derived pyrimidine fused heterocyclic compounds.

and computational studies in these investigations and noted that the efficiency increased with increase in the concentration of compounds. These compounds were reported to be of mixed type with predominantly cathodic type inhibition. Theoretical studies were conducted using quantum chemical calculations and molecular dynamics simulations. It was concluded that the adsorption of these compounds on the metal surface followed both physisorption and chemisorption. Even though theoretical and experimental studies suggested that the efficiency increased in presence of both electron donating as well as

Materials Research Forum LLC
https://doi.org/10.21741/9781644901496-5

electron withdrawing functional groups, authors concluded that compounds with electron donating groups exhibited higher protection than the molecules with electron withdrawing substituents.

Bing Lin and Yu Zuo investigated [28] the inhibition efficiencies of some organic carboxylate compounds having various chain lengths of alkylene (Fig. 12) substitution on Q235 steel with the help of surface analysis, electrochemical measurements and quantum chemical calculations. It was observed that as the alkylene chain length increased, the absolute surface charge value increased, and inhibition efficiency also increased. The best inhibitory effect was exhibited by compound having C_{11} length. Authors examined the effect of acrylic acid, allylacetic acid, 6-heptenoic acid, undecylenic acid and oleic acid. From the results obtained, it was concluded that the adsorption scope of the inhibitor improved with increasing chain length between the C=C bond and the –COO⁻group. Authors suggested that these inhibitors were adsorbed on metal surface by the formation of Fe-OOC-Cx compounds and the presence of C=C bonds enhanced the adsorption process.

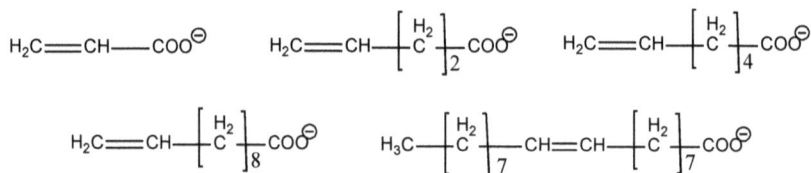

Figure 12. Structure of organic carboxylate compounds having various chain lengths of alkylene substitution.

Feng Yang et al. [29] in their interesting research communication described anti corrosive behaviour of a Zinc – rich epoxy coating containing sulfonated polyaniline (SPANi) in 3.5% NaCl solution on Q235 steel employing electrochemical impedance spectroscopy (EIS) and scanning vibrating electrode technique (SVET). Authors reported that Zinc-rich coating with addition of 1.0% wt SPANi enhanced cathodic protection time and barrier performance.

Wei et al. studied [30] anticorrosive property of 4-phenyl pyrimidine (Fig. 13) coating against copper surface in presence of 3 wt% NaCl utilizing electrochemical impedance spectroscopy (EIS) and polarization methods. Authors reported 83.2% inhibition efficiency. Author used different concentrations of 4-PPM for this purpose like 0.05, 0.10; 0.50 and 1 mm and various immersion times 1, 3, 6 and 10 h respectively. It was concluded that the copper was protected from salt corrosion, by the adsorption of 4PPM molecule on the surface via the N_1 atom in the pyrimidine ring and the formation of uniformly ordered compact layer of 4-PPM molecules.

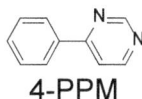

4-PPM

Figure 13.   Structure of 4-phenyl pyrimidine.

Ekemini Akpan and co-workers [31] described corrosion inhibition property of several acridine based thiosemicarbazones (Fig. 14) against mild steel in presence of 1M HCl solution. Corrosion inhibition efficiencies were observed to increase with increase in concentration of each compound. At optimum inhibition concentration the efficiencies are in the order IAB-NF (90.48%) >IAB-ND (87.48%) > IAB-NP (85.28%). Authors reported that these compounds exhibited mixed type behaviour and observed both physisorption and chemisorption mechanisms. Experimental data was supported by both theoretical DFT and Monte Carlo simulation studies.

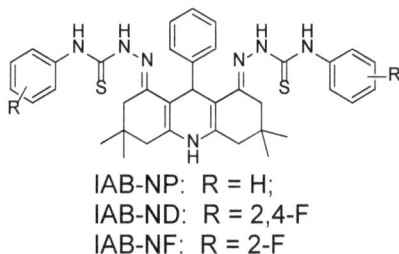

IAB-NP: R = H;
IAB-ND: R = 2,4-F
IAB-NF: R = 2-F

Figure 14.   Structure of acridine based thiosemicarbazones.

Jiyaul Haque et al. [32] synthesized three chitosan Schiff bases (Fig. 15) in an eco-friendly approach under microwave conditions by reacting chitosan and aromatic aldehydes; characterized them spectroscopically and studied their corrosion inhibition behaviour for mild steel in presence of 1 M hydrochloric acid. Authors utilized EIS and PDP techniques for their work. Chitosan Schiff base obtained from 4-hydroxy-3-methoxy benzaldehyde was reported to exhibit maximum inhibition efficiency of 90.65% at a concentration of 50 ppm. They were described as mixed type inhibitors.

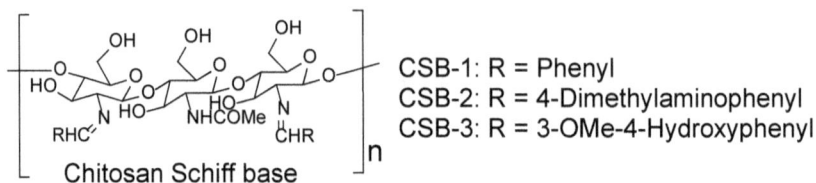

CSB-1: R = Phenyl
CSB-2: R = 4-Dimethylaminophenyl
CSB-3: R = 3-OMe-4-Hydroxyphenyl

Chitosan Schiff base

Figure 15.   Structure of chitosan Schiff bases.

Xiang Gao et al. investigated [33] the corrosion inhibitory efficiencies of three monoalkyl phosphate esters having various chain lengths, such as mono-n-butyl phosphate ester, mono-n-hexylphosphate ester, mono-n-octylphosphate ester, for iron in presence of 0.5 M sulphuric acid solution by using electrochemical impedance spectroscopy (EIS), and polarization curve methods. Authors observed that these behaved as mixed type inhibitors. Authors reported mono-butyl-phosphate ester displayed lower efficiency when compared to the other two esters, whereas mono-n-hexyl as well as mono-n-octyl phosphate esters exhibited almost similar efficiencies. Authors concluded from X-ray photoelectron spectroscopic (XPS) studies that the three alkyl phosphate esters were adsorbed on the metal surface through the unionized P-OH groups.

Alejandro Ramirez-Estrada et al. [34] developed environmentally benign synthetic zwitterionic compounds derived from β-amino acids prepared from alkylamines such as octylamine, dodecylamine and octadecylamine and acrylic acid under solvent free and ambient conditions (Fig. 16). After characterization, these compounds were evaluated for their anticorrosion efficiency for carbon-steel under CO_2 acidic environments characteristic of oil- field environments with the help of electrochemical techniques such as open circuit potential (OCP), linear polarization potential (LRP), and electrochemical impedance spectroscopy (EIS). These compounds exhibited 60%, 89% and 90% inhibition efficiencies respectively for the compounds obtained from octylamine, dodecylamine and octadecylamine (GZC-8; GZC-12; GZC-18), respectively at a concentration of 25 ppm. Authors observed that these were of mixed type of inhibitors and described that GZC-18 showed best efficiency of 90.14%. Authors claim that this is the first report on the preparation and evaluation of nontoxic zwitterionic corrosion inhibitors as alternative to typical amphiphilic and ionic liquid types and offered beginning for the GCZ compounds as green corrosion inhibitors for different industries. Authors also conclude that long hydrophobic substitution on an amphiphilic molecule augments corrosion protection when compared to a molecule with smaller hydrophobic substituents. Authors also opined that GZC-12 was the best option with higher efficiency and lower toxicity as observed from the experimental data.

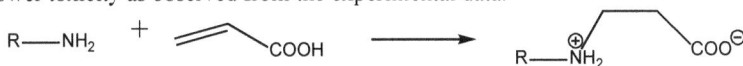

GZC-8: = R = n-Octyl
GZC-12: = R = n-Dodecyl
GZC-18: = R = n-Octadecyl

Figure 16.   Structure of zwitterionic compounds derived from β-amino acids prepared from alkylamines such as octylamine, dodecylamine and octadecylamine and acrylic acid.

Nicola Weder and co-authors [35] investigated the adsorption and corrosion inhibition mechanisms of some different symmetrical N,N`-thiourea derivatives (Fig. 17) on aluminium in hydrochloric acid. Authors observed that the corrosion rate measurements indicated cathodic type inhibition by all compounds. Authors selected thiourea, dimethyl thiourea, diethyl thiourea, dipropylthiourea, dibutylthiourea, diphenylthiourea, tetramethylthiourea etc., for their study.

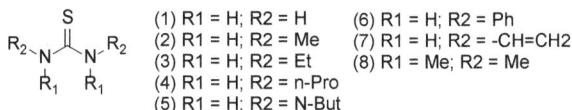

(1) R1 = H; R2 = H   (6) R1 = H; R2 = Ph
(2) R1 = H; R2 = Me   (7) R1 = H; R2 = -CH=CH2
(3) R1 = H; R2 = Et   (8) R1 = Me; R2 = Me
(4) R1 = H; R2 = n-Pro
(5) R1 = H; R2 = N-But

Figure 17.   Structure of symmetrical N,N`-thiourea derivatives.

Chandrabhan Verma et al. examined [36] corrosion inhibition capacity of three N-substituted 2-aminopyridine derivatives (Fig. 18) on mild steel surfaces in 1 M HCl solution employing theoretical, chemical, surface, and electrochemical studies. It was reported that the efficiency increased with increase in the concentration of the chemical constituent and reached maximum at 20.20×10^{-5} molL^{-1} concentration. Inhibition efficiency was observed to be in the order of DMPN > DHPN > DPPN. Authors noted

Sustainable Corrosion Inhibitors   Materials Research Forum LLC
Materials Research Foundations107 (2021) 101-129 https://doi.org/10.21741/9781644901496-5

that these belong to mixed type inhibitors with predominant cathodic inhibitive action. Experimental studies indicated that these molecules were adsorbed on metal surface as protonated species.

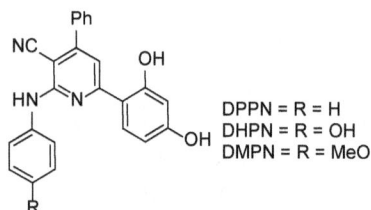

Figure 18. Structure of three N-substituted 2-aminopyridine derivatives.

Cano et al. studied [37] Fast green (FG), fuchsine base (FB) and fuchsine acid (FA) (Fig. 19) for copper corrosion inhibition in presence of citric acid solution of concentrations ranging from 0.001 to 1.0 M. Authors observed inhibition efficiency in the order FB > FA > FG.

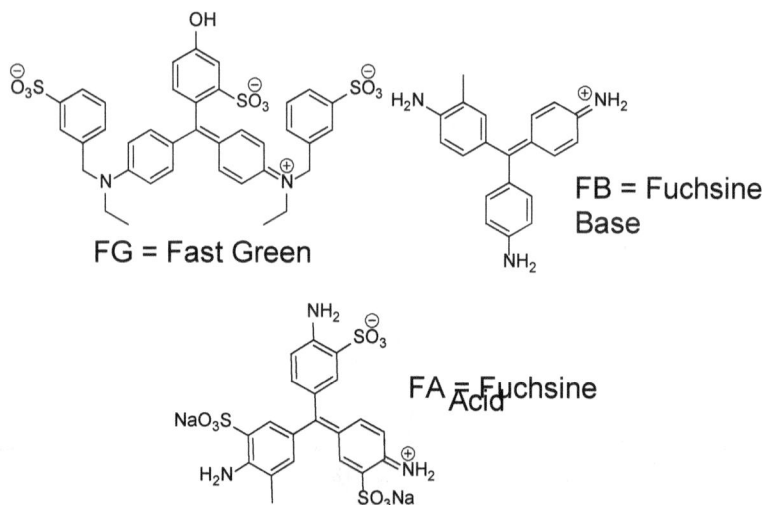

Figure 19. Structure of Fast green (FG), fuchsine base (FB) and fuchsine acid (FA).

Saad Kaskah et al. [38] investigated anticorrosion properties of N-acyl sacrosine derivatives (Fig. 20) as green corrosion inhibitors for low carbon steel. Authors selected eco-friendly fatty acid N-acyl derivatives of the naturally occurring amino acid sarcosine such as lauroyl, myristoyl and oleoylsarcosine for their study on two low carbon steel variants for the comparison. Both steel variants comply with the standards but differ in the long-term anticorrosion behaviour. Authors utilized electrochemical impedance spectroscopy and potentiodynamic polarization techniques and concluded that the inhibitors belonged to the mixed type. Best results were obtained with oleoylsarcosine (C-18) with 99.6% and 97.4% for the two different steel typesfollowed by myristoyl (C-14) and lauroyl (C-12) sarcosine derivatives. The trend of efficiencies observed is correlated with the chain length of the hydrophobic part. Oleoylsarcosine (C-18) with longest alkyl side chain having additional double bond showed maximum efficiency.

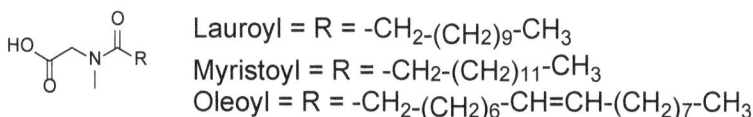

Lauroyl = R = $-CH_2-(CH_2)_9-CH_3$
Myristoyl = R = $-CH_2-(CH_2)_{11}-CH_3$
Oleoyl = R = $-CH_2-(CH_2)_6-CH=CH-(CH_2)_7-CH_3$

Figure 20.   Structure of N-acyl sacrosine derivatives.

Roghayeh Sadeghi Erami and co-workers [39] described anti corrosion properties exhibited by carboxamide ligands (N-(quinolin-8-yl)quinoline-2-carboxamide (Hqcq) and N-(quinolin-7-yl)pyrazine-2-carboxamide (Hqpzc) (Fig. 21)in presence of 1 M HCl solution at 25 °C for mild steel surfaces. Both the compounds were synthesized in an eco-friendly manner using ionic liquid TBAB. Due to uniform layer formation and plannar and parallel arrangement on the metal surface Hqcq showed better anti corrosion property. Experimental studies indicated that both the compounds exhibited mixed type of mechanism tending more towards chemisorption process.

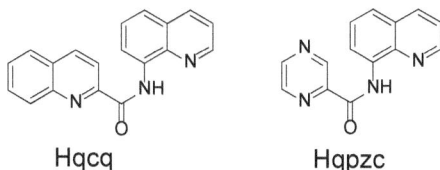

Hqcq   Hqpzc

Figure 21.   Structure of carboxamide ligands (N-(quinolin-8-yl) quinoline-2-carboxamide (Hqcq) and N-(quinolin-7-yl) pyrazine-2-carboxamide (Hqpzc).

Ikenna et al. [40] assessed 2-(2-pyridyl)benzimidazole(PB) (Fig. 22) as a promising green corrosion inhibitor for APIX60 steel for oil-field application, in sweet corrosive environment and observed 2-PB to be more effective in hydrodynamics than in static conditions, where in mass transfer of HCO_3^- species was protonated facilitating the precipitation of Fe_2CO_3 resulting in corrosion resistance. Authors utilized electrochemical techniques such as electrochemical impedance spectroscopy (EIS), linear polarization resistance (LPR) and potentiodynamic polarization (PDP), in their studies. It is suggested that 2-PB get adsorbed on the steel surface with the help of NH-group of benzimidazole skeleton.

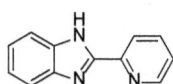

Benz-imidazole Skeleton

Figure 22.   Structure of 2-(2-pyridyl)benzimidazole(PB).

Dipankar Sukul and coworkers [41] evaluated some newly synthesized quercetin derivatives (Fig. 23) as corrosion inhibitors for mild steel in presence of 1 M HCl solution backed by both experimental and theoretical studies. Quercetin, a poly phenol and a plant derived aglycone form of Rutin, a flavonoid glycoside, exhibits high antioxidant property. Authors prepared 6-((4-(2-hydroxyethy)piperazine-1-yl)methyl)-quercetin (A), and 6,8-di-((4-(2-hydroxyethy)piperazine-1-yl)methyl)-quercetin(B) in order to enhance water solubility. A and B are made from paraformaldehyde, 2-piperazinyl ethanol, and quercetindihydrate.

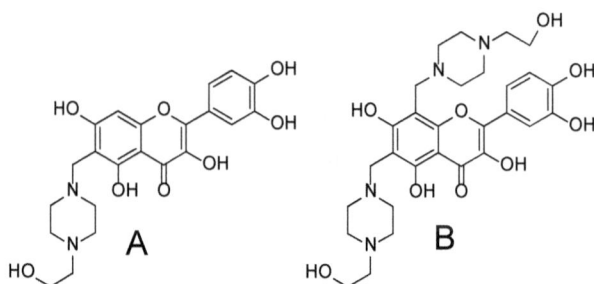

Figure 23.   Structure of quercetin derivatives.

Authors utilized electrochemical potentiodynamic polarization, wight loss studies, electrochemical impedance spectroscopy and SEM images. Both compounds exhibited above 90% inhibition efficiency towards corrosion of mild steel in 1 M HCl from ambient to 323 K temperature at 1mM concentration level: and also retarded the rates of both cathodic and anodic reactions by blocking the surface reactive sites.

Yadav and co-workers [42] described the synthesis of two spiroindolinopyrimidine derivatives (1) MPTS (2) CPTS (Fig. 24) following the previously reported experimental procedure and assessed their effect on the corrosion of mild steel in 15% hydrochloric acid. Data suggested that both the compounds behaved as mixed type inhibitors and the surface morphology was analyzed with the help of SEM, XPS, FTIR, AFM and energy dispersion X-ray spectroscopy. Effect of the molecular structure on the inhibition efficiency was studied using quantum chemical calculations with the help of DFT. Molecule MPTS exhibited better performance when compared to CPTS. It was concluded that the corrosion inhibition mechanism was due to the adsorption of spiroindolinopyrimidine molecules on the metal surface. In this research work authors also corroborated experimental results with theoretical studies.

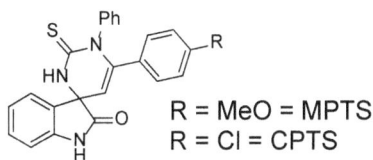

Figure 24.   Structure of spiroindolinopyrimidine derivatives (1) MPTS (2) CPTS.

8-Hydroxyquinoline was examined as a new generation corrosion inhibitor by Fatah Chiter et al. [43]. Authors investigated the preference of 8-hydroxyquinoline and its compounds on the Al(III) surface and comprehensive DFT studies were undertaken. Authors assessed the formation of layers of 8-hydroxyquinoline and derivatives as well as the influence of the coverage of Al(III) surface on the adsorption topologies and adsorption energies of organic molecules, charge transfer due to the adsorption process, and the interface dipoles created at the molecule/metal interface. Authors demonstrated that among 8HQ and its derivatives, dehydrogenated 8HQ molecules formed the most stable layer with strongest coupling with the Al(III) surface, with a topology of the molecules tilted on the metallic surface.

Singh and co-workers [44] reported green microwave assisted synthesis of 2-amino-N-(thiophen-2yl)methylene benzohydrazide (ATMBH) (Fig. 25) under solvent free conditions in presence of montmorillonite clay, an environmentally benign solid acid catalyst, characterized its structure and examined anticorrosion property against mild steel in presence of 0.5 M sulphuric acid utilizing gravimetric and electrochemical methods. Authors described ATMBH to be more effective on cathodic branch of corrosion process, chemical adsorption being more favoured at higher concentrations.

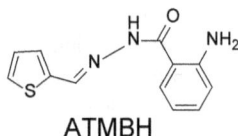

ATMBH

Figure 25.   Structure of 2-amino-N-(thiophen-2yl)methylene benzohydrazide (ATMBH).

Neeraj Kumar Gupta and co-workers [45] synthesized three cysteine based Schiff bases CSB1, CSB2, CSB3 (Fig. 26) and evaluated their corrosion inhibition efficiency for mild steel surfaces in presence of 1 M hydrochloric acid employing different scientific studies. Authors prepared these different title compounds by reacting cysteine and various aldehydes such as benzaldehyde, 4-methoxy benzaldehyde, and 3-methoxy-4- hydroxy benzaldehyde employing known eco-friendly methodologies. These exhibited both physisorption and chemisorption and among all the three compounds CSB3 exhibited better efficiency with 97.3% at 200 ppm concentration. From EIS study results, authors concluded that these compounds inhibited corrosion by increase of the charge transfer resistance between metal-solution interfaces.

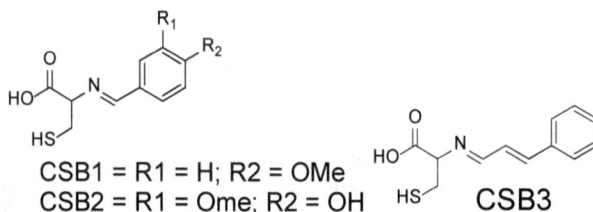

CSB1 = R1 = H; R2 = OMe
CSB2 = R1 = Ome; R2 = OH
CSB3

Figure 26.   Structure of cysteine-based Schiff bases CSB1, CSB2, CSB3.

Chandrabhan Verma and co-authors [46] synthesized three new 2,4-diamino-5-(phenylthio)-5H-chromeno[2,3-b]-pyridine-3-carbonitriles (DHPCS) by MCR (Fig. 27) and investigated the anticorrosion properties of the three compounds for mild steel in

Materials Research Forum LLC
https://doi.org/10.21741/9781644901496-5

presence of 1 M HCl with weight loss method, electrochemical techniques and surface morphology studies (SEM, AFM) supported by theoretical methods (quantum chemical calculations and molecular dynamics studies). Based on the data gathered, authors informed that the inhibition efficiency increased with increasing concentration of inhibitor and concluded that the efficiency of three compounds followed trend DHPC-3 > DHPC-2 > DHPC-1. All these acted as mixed type inhibitors, exhibiting both physisorption and chemisorption. It was considered that the presence of electron withdrawing group was responsible for lowest inhibitive efficiency for DHPC-1. Furthermore, DHPC-3 showed the highest efficiency, as concluded from the observations, due to the presence of electron releasing-OH group at position seven of chromenopyridine ring.

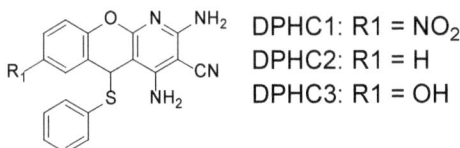

DPHC1: R1 = NO_2
DPHC2: R1 = H
DPHC3: R1 = OH

Figure 27.   Structure of 2,4-diamino-5-(phenylthio)-5H-chromeno[2,3-b]-pyridine-3-carbonitriles (DHPCS).

β-Sitosterol (Fig. 28) was isolated from rice hulls by Maya Krishnan Prabhakaran and co-workers [47] and its corrosion inhibition efficiency was assessed for mild steel surface in presence of 1 M sulphuric acid solution. Authors employed FTIR spectroscopy, weight loss measurements, electrochemical studies and atomic absorption spectroscopy to examine corrosion inhibition behaviour of sitosterol and observed that inhibition efficiency increased with increase in inhibitor concentration and decreased with increase in temperature. β-Sitosterol exhibited maximum inhibition efficiency of 95% at 500 ppm level. Authors inform that efficiency of β-sitosterol may be due to the relatively flat structure of molecule with hydrophobicity. β-Sitosterol acted as mixed type inhibitor controlling both anodic and cathodic processes.

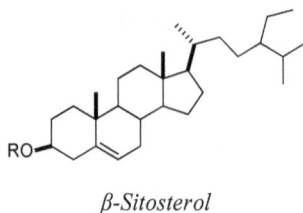

β-Sitosterol

Figure 28.   Structure of β-Sitosterol.

Zheludkevich et al. [48] developed a bi-layer protective coating based on chitosan and examined for application for aluminium alloy 2024, that had wider applications. This new type of corrosion protecting self healing coating consists of a chitosan-based pre-layer coated on the metal surface along with a barrier hybrid layer. The chitosan layer was doped with cerium ions to serve as a pool for corrosion inhibitor. The complex formed by Cerium ion and chitosan functional groups provide for a continuous release of the active agent. Authors demonstrated superior corrosion protection with the Cerium doped biopolymer pre-layer. Chitosan is a linear polysaccharide consisted of units of 2-amino-2-deoxy-D-glucose. It has nice layer forming property with superior adhesion to metallic surfaces and to many organic polymers. It has applications in several fields like pharmaceuticals, biomedicine, cosmetics, biotechnology and in protective coatings. About four types of coatings were studied by authors, an un-doped chitosan film, chitosan film doped with Cerium nitrate, an undoped chitosan film over coated with sol-gel and Cerium containing chitosan layer covered with sol-gel. Chitosan films exhibit excellent adhesion to both metallic surface and the hybrid sol-gel coating. Cerium ions inhibited corrosion forming hydroxide deposits atop the active intermetallic, blocking electrochemical activity, providing superior corrosion protection. Moreover, the released Cerium cations are expected to actively suppress the corrosion process by stabilizing the metal-polymer interface. Authors concluded that the developed self healing coating system can be assumed to be a favourable green solution for corrosion inhibition.

Jiyaul Haque et al. [49] investigated N-methyl-N,N,N-triacetyl ammoniumchloride (Aliquat 336) (Fig. 29) as a novel and green corrosion inhibitor for mild steel in acid chloride medium employing potentiodynamic polarization (PDP), other spectroscopic methods (FTIR, UV-Vis) and also theoretical methods. Aliquat 336 exhibited good efficiency of 94.6%. Authors showed the title compound as a mixed type inhibitor. Aliquat 336 has high hydrophobic character because of the presence of three long

Materials Research Forum LLC
https://doi.org/10.21741/9781644901496-5

alkylchain substituents and a quaternary N atom and has wider applications as a phase transfer catalyst, metal extracting agent and surfactant.

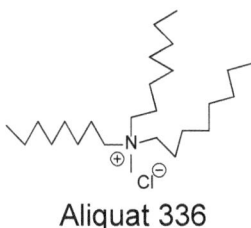

Aliquat 336

Figure 29.   Structure of N-methyl-N,N,N-triacetyl ammoniumchloride (Aliquat 336).

Yadav and co-workers [50] synthesized two carbohydrate compounds (Fig. 30) and examined their corrosion inhibition and adsorption characteristics on oil-field N-80 steel in presence of 15% HCl solution using EIS, PZC, Polarization and mass loss procedures. Experimental data was corroborated with DFT and molecular dynamics simulations (MDS). Authors reported 98.3% and 91.5% inhibitor efficiency for BIHT and MIHT respectively at 333 K, in presence of 400 ppm level. It is concluded that BIHT binds stronger on mild steel surface than MIHT due to the presence of two extra benzene rings associated with BIHT structure.

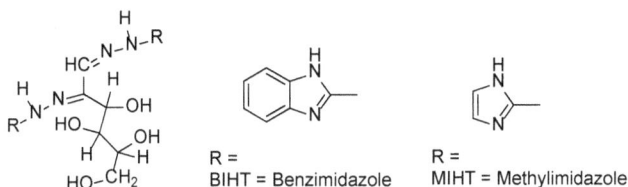

Figure 30.   Structure of BIHT and MIHT; carbohydrate compounds.

Conclusion

The last few decades have witnessed substantial growth in finding new organic corrosion inhibitors. The present review was authored purely out of academic interest to familiarize the young science researchers to appraise them about the recent work appearing on green organic corrosion inhibitors, particularly related to organic molecules. The examples covered in this review are chosen from different journals. The authors of this review are highly appreciative of the research articles published for their contributions in the area of organic corrosion inhibitors. This review is only representative in nature and not intended to be exhaustive. The readers are advised to go through original research communications for detailed information. The figures are drawn simply and selectively in a representative manner.

Acknowledgement

The authors of this review acknowledge the original contributors and publishers of the articles cited here for their potential scientific work.

VJR thanks Dr B.Parthasaradhy Reddy, Chairman Heterodrugs, Pvt. Ltd. and Dr K. Ratnakar Reddy, Director HR Foundation for their encouragement. VJR also thanks CSIR New Delhi for Emeritus Scientist honour.

References

[1] Rodríguez-Torres, O. Olivares-Xometl, M.G. Valladares-Cisneros, J.G. González-Rodríguez, Effect of Green Corrosion Inhibition by Prunuspersica on AISI 1018 Carbon Steel in 0.5M H2SO4, Int. J. Electrochem. Sci. 13 (2018) 3023-3049. https://doi.org/10.20964/2018.03.40

[2] G. Subramanian, R.S. Kannan, M. Malarvizhi, P. Muthirulan, A novel eco-friendly corrosion inhibitor for mild steel protection in two different aggressive artificial corrosive medium, J. Chem. Pharm. Res. 10 (2018) 155-163.

[3] L.A.L. Guedes, K.G. Bacca, N.F. Lopes, E.M. da Costa, Tannin of Acacia mearnsiias green corrosion inhibitor for AA7075-T6 aluminium alloy in acidic medium, Mater. Corros. 70 (2019) 1288-1297. https://doi.org/10.1002/maco.201810667

[4] H. Taoui, H. Bentrah, A. Chala, M. Djellab, Bark resin of Schinusmolle as an eco-friendly inhibitor for API 5L X70 pipeline steel in HCl medium, Mater. Corros. 70 (2019) 511-520. https://doi.org/10.1002/maco.201810477

[5] A.S. Yaro, A.A. Khadom, R.K. Wael, Apricot juice as green corrosion inhibitor of mild steel in phosphoric acid, Alexandrai Eng. J. 52 (2013) 129-135. https://doi.org/10.1016/j.aej.2012.11.001

[6] Aiad, S.M. Shaban, A.H. Elged, O.H. Aljoboury, Cationic surfactant based on alignate as green corrosion inhibitors for the mild steel in 1.0 M HCl, Egypt. J. Petrol. 27 (2018) 877-885. https://doi.org/10.1016/j.ejpe.2018.01.003

[7] M.T. Saeed, M. Saleem, S. Usmani, I.A. Malik, F.A. Al-Shammari, K.M. Deen, Corrosion inhibition of mild steel in 1 M HCl by sweet melon peel extract, J. King Saud Univ. Sci. 31 (2019) 1344-1351. https://doi.org/10.1016/j.jksus.2019.01.013

[8] A.A. Khadom, A.N. Abd, N.A. Ahmed, Xanthium Strumarium leaves extracts as a friendly corrosion inhibitor of low carbon steel in hydrochloric acid: kinetics and mathematical studies; South Afr. J. Chem. Eng. 25 (2018) 13-21. https://doi.org/10.1016/j.sajce.2017.11.002

[9] N. Raghavendra, J.I Bhat, Red arecanut seed extract as a sustainable corrosion inhibitor for aluminum submerged in acidic corrodent: An experimental approach towards zero environmental impact, Periodica. Polytechnica. Chem. Eng. 62 (2018) 351-358. https://doi.org/10.3311/PPch.10686

[10] I. Pradipta, D. Kong, J. Ban, L. Tan, Natural organic antioxidants from green tea form a protective layer to inhibit corrosion of steel reinforcing bars embedded in mortar, Construct. Build. Mater. 221 (2019) 351-362. https://doi.org/10.1016/j.conbuildmat.2019.06.006

[11] R. Idouhli, Y. Koumya, M. Khadiri, A. Aityoub, A. Abouelfda, A. Benyaich, Inhibitory effect of Senecioanteuphorbium as green corrosion inhibitor for S300 steel, Int. J. Ind. Chem. 10 (2019) 133-143. https://doi.org/10.1007/s40090-019-0179-2

[12] A. Rodríguez-Torres, M.G. Valladares-Cisneros, C. Cuevas-Arteaga, M.A. Veloz-Rodríguez, Study of green corrosion inhibition on AISI 1018 carbon steel in sulfuric acid using crataegusmexicana as eco-friendly inhibitor. J. Mater. Environ. Sci. 10 (2019) 101-112.

[13] S. Umoren, Z.M. Gasem, Ime B Obot, Natural products for materials protection: Inhibition of mild steel corrosion by date palm seed extracts in acid media, Ind. Eng. Chem. Res. 52 (2013) 14855-148655. https://doi.org/10.1021/ie401737u

[14] P.B. Raja, M. Fadaeinasab, A.K. Qureshi, A.A. Rahim, H. Osman, M. Litaudon, K. Awang, Evaluation of green corrosion inhibition by alkaloid extracts of Ochrosiaoppositifolia and isoreserpiline against mild steel in 1M HCl medium, Ind. Eng. Chem. 52 (2013) 10582-10593. https://doi.org/10.1021/ie401387s

[15] N.I.N. Haris, S. Sobri, N. Kassim, Oil palm empty fruit bunch extract as green corrosion inhibitor for mild steel in hydrochloric acid solution: Central composite design optimization, Mater. Corros. 70 (2019) 1111-1119. https://doi.org/10.1002/maco.201810653

[16] R. Haldhar, D. Prasad, A. Saxena1, P. Singh, valerianawillichiroots extract as a green & sustainable corrosion inhibitor for mild steel in acidic environment: experimental and theoretical study, Mater. Chem. Frontiers 2 (2018) 1225-1237. https://doi.org/10.1039/C8QM00120K

[17] K. Zhang, W. Yang, B. Xu, X. Yin, Y. Chen, Y. Liu, Green synthesis of novel schiff bases as eco-friendly corrosion inhibitors for mild steel in hydrochloric acid, Chem. Select 3 (2018) 12486-12494. https://doi.org/10.1002/slct.201802915

[18] O. Kaczerewska1, R. Leiva-Garcia, R. Akid, B. Brycki, I. Kowalczyk, T. Pospieszny, Heteroatoms and π electrons as favorable factors for efficient corrosion protection, Mater. Corros. 70 (2019) 1099-1110. https://doi.org/10.1002/maco.201810570

[19] S. Varvara, R. Bostan, L. Gaina, L.M. Muresan, Thiadiazole derivatives as inhibitors for acidic media corrosion of artificially patinated bronze, Mater. Corros. 65 (2014) 1202-1214. https://doi.org/10.1002/maco.201307072

[20] T.M.A. Eldebss, A.M. Farag, A.Y.M. Shamy, Synthesis of some benzimidazole-based heterocycles and their application as copper corrosion inhibitors, J. Heterocyclic Chem. 56 (2019) 371-390. https://doi.org/10.1002/jhet.3407

[21] W. Zhang, H.J. Li, A. Wang, C. Ma, Z. Wang, H. Zhang, Y.C. Wu, Synergistic inhibition effect of N-(furan-2-ylmethyl)-7 H-purin-6-amine and iodide ion for mild steel corrosion in 1 mol/L HCl, Mater. Corros. 70 (2019) 1-10. https://doi.org/10.1002/maco.201911146

[22] E.A. Yaqo, R.A. Anaee, M.H. Abdulmajeed, I.H.R. Tomi, M.M. Kadhim, Aminotriazole derivative as anti-corrosion material for iraqi kerosene tanks: Electrochemical, computational and the surface study, Chem. Select 4 (2019) 9883-9892. https://doi.org/10.1002/slct.201902398

[23] H. Liu, J. Hu, X. Zhou, D. Liu, B. Xu, Y. Zhou, Synthesis, corrosion inhibition performance and biodegradability of novel alkyl hydroxyethylimidazoline salts, J. Surface Deteg. 18 (2015) 1025-1031. https://doi.org/10.1007/s11743-015-1725-3

[24] T.K.A. Hoang, T.N.L. Doan, J.H. Cho, J.Y.J. Su, C. Lee, C. Lu, P. Chen, Sustainable gel electrolyte containing pyrazole as corrosion inhibitor and dendrite suppressor for aqueous Zn/LiMn2O4 battery, Chem. Sus. Chem. 10 (2017) 2816-2822. https://doi.org/10.1002/cssc.201700441

[25] I. Aiad, S.M. Shaban, A.H. Elged, O.H. Aljoboury, Cationic surfactant based on alignate as green corrosion inhibitors for the mild steel in 1.0 M HCl, Egypt. J. Petrol. 27 (2018) 877-885. https://doi.org/10.1016/j.ejpe.2018.01.003

[26] P. Dohare, M.A. Quraishi, C. Vermac, H. Lgaze, R. Salghif, E.E. Ebenso, Ultrasound induced green synthesis of pyrazolo-pyridines as novel corrosion inhibitors useful for industrial pickling process: Experimental and theoretical approach, Results Phys. 13 (2019) 102344. https://doi.org/10.1016/j.rinp.2019.102344

[27] Verma, E.E. Ebenso, L.O. Olasunkanmi, M.A. Quraishi, I.B. Obot, Adsorption behavior of glucosamine based pyrimidinefusedheterocycles as green corrosion inhibitors for mild steel: Experimental and theoretical studies, J. Phys. Chem. C, 120 (2016) 11593-11611. https://doi.org/10.1021/acs.jpcc.6b04429

[28] Lin, Y. Zuo, Corrosion inhibition of carboxylate inhibitors with different alkylene chain lengths on carbon steel in an alkaline solution, RSC Adv. 9 (2019) 7065-7077. https://doi.org/10.1039/C8RA10083G

[29] F. Yang, T. Liu, J. Li, S. Qiu, H. Zhao, Anticorrosive behavior of a zinc-rich epoxy coating containing sulfonatedpolyaniline in 3.5% NaCl solution, RSC Adv. 8 (2018) 13237-13247. https://doi.org/10.1039/C8RA00845K

[30] N. Wei, Y. Jiang, Z. Liu, Y. Ying, X. Guo, Y. Wu, Y. Wen, H. Yang, 4-Phenylpyrimidine monolayer protection of a copper surface from salt corrosion; RSC Adv. 8 (2018) 7340-7349. https://doi.org/10.1039/C7RA12256J

[31] E.D. Akpan, I.O. Isaac, L.O. Olasunkanmi, E.E. Ebenso, E.S.M. Sherif, Acridine-based thiosemicarbazones as novel inhibitors of mild steel corrosion in 1 M HCl: Synthesis, electrochemical, DFT and Monte Carlo simulation studies, RSC Adv., 9 (2019) 29590-29599. https://doi.org/10.1039/C9RA04778F

[32] J. Haque, V. Srivastava, D.S. Chauhan, H. Lgaz, M.A. Quraishi, Microwave-induced synthesis of chitosan schiff bases and their application as novel and green corrosion inhibitors: Experimental and theoretical approach, ACS Omega 3 (2018) 5654-5668. https://doi.org/10.1021/acsomega.8b00455

[33] X. Gao, S. Liu, H.F. Lu, F. Gao, H. Ma, Corrosion inhibition of iron in acidic solutions by monoalkyl phosphate esters with different chain lengths, Ind. Eng. Chem. 54 (2015) 1941-1952. https://doi.org/10.1021/ie503508h

[34] A.R. Estrada, V.Y.M. Cervantes, I. Elizalde, A.M. Robledo, L.S.Z. Rivera, D.A.N. Álvarez, F. Farelas, R.H. Altamirano, Development of a zwitterionic compound derived from β-amino acid as a green inhibitor for co2 corrosive environments; ACS

Sustain. Chem. Eng. 5 (2017) 10396-10406.
https://doi.org/10.1021/acssuschemeng.7b02434

[35] N. Weder, R.A. Alberto, R. Koitz, Thiourea derivatives as potent inhibitors of aluminium corrosion: Atomic-level insight into adsorption and inhibition mechanisms, J. Phys. Chem. C 120 (2016) 1770-1777. https://doi.org/10.1021/acs.jpcc.5b11750

[36] P.B. Raja, M. Fadaeinasab, A.K. Qureshi, A.A. Rahim, H. Osman, M. Litaudon, K. Awang, Evaluation of green corrosion inhibition by alkaloid extracts of Ochrosiaoppositifolia and isoreserpiline against mild steel in 1M HCl medium, Ind. Eng. Chem., 52 (2013) 10582-10593. https://doi.org/10.1021/ie401387s

[37] Cano, P. Pinilla, J.L. Polo, J.M. Bastidas, Copper corrosion inhibition by fast green, fuchsin acid and basic compounds in citric acid solution, Mater. Corros. 54 (2003) 222-228. https://doi.org/10.1002/maco.200390050

[38] S.E. Kaskah, M. Pfeiffer, H. Klock, H. Bergen, G. Ehrenhaft, P.F.J. Gollnick, C.B. Fischer, Surface protection of low carbon steel with N-acyl sarcosine derivatives as green corrosion inhibitors; Surfaces and Interfaces, 9 (2017) 70-78. https://doi.org/10.1016/j.surfin.2017.08.002

[39] R.S. Erami, M. Amirnasr, S. Meghdadi, M. Talebian, H. Farrokhpour, K. Raeissi, Carboxamide derivatives as new corrosion inhibitors for mild steel protection in hydrochloric acid solution, Corros. Sci. 151 (2019) 190-197. https://doi.org/10.1016/j.corsci.2019.02.019

[40] I.B. Onyeachu, I.B. Obot, A.A. Sorour, M.I.A. Rashid, Green corrosion inhibitor for oilfield application I: Electrochemical assessment of 2-(2-pyridyl) benzimidazole for API X60 Steel under sweet environment in NACE brine ID196, Corros. Sci. 150 (2019) 183-193. https://doi.org/10.1016/j.corsci.2019.02.010

[41] Sukul, A. Pal, S.K. Saha, S. Satpati, U. Adhikari, P. Banerjee, Newly synthesized quercetin derivatives as corrosion inhibitor for mild steel in 1 M HCl: Combined experimental and theoretical investigation, Phys. Chem. Chem. Phys. 20 (2018) 6562-6574. https://doi.org/10.1039/C7CP06848D

[42] M. Yadav, R.R. Sinha, S. Kumar, T.K. Sarkar, Corrosion inhibition effect of spiropyrimidinethiones on mild steel in 15% HCl solution insight from: Electrochemical and quantum studies, RSC Adv. 5 (2015) 70832-70848. https://doi.org/10.1039/C5RA14406J

[43] Chiter, M.L. Bonnet, C.L. Dufaure, H. Tangb, N. Pe'be're, Corrosion protection of Al(111) by 8-hydroxyquinoline: a comprehensive DFT study, Phys. Chem. Chem. Phys. 20 (2018) 21474-21486. https://doi.org/10.1039/C8CP03312A

[44] A.K. Singh, S. Thakur, B. Pani, G. Singh, Green synthesis and corrosion inhibition study of 2-amino-N'-(thiophen-2-yl) methylene) benzohydrazide, New J. Chem. 42 (2018) 2113-2124. https://doi.org/10.1039/C7NJ04162D

[45] N.K. Gupta, M.A. Quraishi, C. Verma, A.K. Mukherjee, Green Schiff's bases as corrosion inhibitors for mild steel in 1 M HCl solution: Experimental and Theoretical approach, RSC Adv. 6 (2016) 102076-102087. https://doi.org/10.1039/C6RA22116E

[46] C. Verma, L.O. Olasunkanmi, I. B. Obot, E.E. Ebenso, M.A. Quraishi, 2, 4-diamino-5-(phenylthio)-5H-chromeno [2, 3-b] pyridine-3-carbonitriles as green and effective corrosion inhibitors: Gravimetric, electrochemical, surface morphology and theoretical studies, RSC Adv. 6 (2016) 53933-53948. https://doi.org/10.1039/C6RA04900A

[47] M. Prabakaran, S.H. Kim, A. Sasireka, V. Hemapriya, I.M. Chung, β-Sitosterol isolated from rice hulls as an efficient corrosion inhibitor for mild steel in acidic environments, New J. Chem. 41 (2017) 3900-3907. https://doi.org/10.1039/C6NJ03760G

[48] M.L. Zheludkevich, J. Tedim, C.S.R. Freire, S.C.M. Fernandes, S. Kallip, A. Lisenkov, A. Gandini, M.G.S. Ferreira, Self-healing protective coatings with "green" chitosan based pre-layer reservoir of corrosion inhibitor, J. Mater. Chem. 21 (2011) 4805-4812. https://doi.org/10.1039/c1jm10304k

[49] J. Haque, V. Srivastava, C. Verma, H. Lgaz, R. Salghi, M.A. Quraishi, N-Methyl-N, N, N-trioctylammonium chloride as novel and green corrosion inhibitor for mild steel in acid chloride medium: Electrochemical, DFT and MD studies, New J. Chem. 41 (2017) 13647-13662. https://doi.org/10.1039/C7NJ02254A

[50] M. Yadav, T.K. Sarkar1, I.B. Obot, Carbohydrate compounds as green corrosion inhibitor: Electrochemical, XPS, DFT and molecular dynamics simulation studies, RSC Adv. 6 (2016), 110053-110069. https://doi.org/10.1039/C6RA24026G

Sustainable Corrosion Inhibitors
Materials Research Foundations107 (2021) 130-146

Materials Research Forum LLC
https://doi.org/10.21741/9781644901496-6

Chapter 6

Sustainable Corrosion Inhibitors for Concrete

B. Chugh[1], S. Thakur[1], B. Pani[2], A.K. Singh[3,*]

[1]Department of Chemistry, Netaji Subhas University of Technology (formerly known as Netaji Subhas Institute of Technology, University of Delhi), New Delhi-110078, India

[2]Department of Chemistry, Bhaskaracharya College of Applied Science, University Of Delhi, New Delhi-110078, India

[3]Department of Applied Sciences, Bharati Vidyapeeth's College of Engineering, Guru Govind Singh Indraprastha University, New Delhi-110063, India

*ashish.singh.rs.apc@itbhu.ac.in

Abstract

Green chemistry and sustainability encourages the significance of preserving the nature and individual's wellbeing in cost effective way that intends to avoid toxicity and reduction of wastes. Therefore, the implication of green corrosion inhibitors in the field of concrete protection has also received immense attention these days. Indeed, the usage of such inhibitors is a well-known strategy for producing high performance concrete. In view of this, the present chapter discusses the research in the area of sustainable corrosion inhibitors for concrete assurance used commercially in various industries. It also highlights the concrete corrosion mechanisms and its protective measures, recent advances in this field.

Keywords

Concrete, Inhibitors, Reinforcement, SEM, Sustainable, EIS

Contents

1.   Introduction

Concrete is a mixture which usually consists of cement, water and aggregates that has been utilised in the major amount as structural material in past years. Cement is one of the primary segment of concrete which produces a paste when dissolved in water that settles and stiffens because of hydration phenomenon. Generally, concrete is relatively feeble in tension, thus changes must be done for the improvement in the tensile strength of the framework which could be by shifting to the different component that is strong in tension. In view of this, concrete frameworks are frequently toughened by embedding steel ribs, usually called as reinforcement in concrete. But, the primary obstacle in reinforcement in concrete is corrosion. Corrosion is a process which brings about the deterioration or decay of a material when exposed to severe environmental conditions [1]. Usually, concrete results in exquisite corrosion shielding. The immense alkaline condition in concrete leads to the development of a firmly adhering film, that passivates the steel, thereby safeguarding it from corrosion [2]. As per the literature, it is expected that the alkalinity of the concrete with pH values ranging from 12 to 14 promotes the production of a passive layer that helps in the shielding of the steel reinforced corrosion [3,4]. Passivation implies the inability of metals to lose electrons in the anodic process that makes metal inert in any corrosive condition. But, the assurance given by the concrete is not adequate because of the porosity of the material and inbuilt crevices which subsequently permits the penetration of aggressive substances like chloride ions and carbon dioxide (CO_2) that results in the reinforcement of steel corrosion by eliminating the protective adhering film over the surface [5]. Rapid degradation could be brought about via chloride ions in the marine region [6] or because of the usage of defrost salts or

Sustainable Corrosion Inhibitors   Materials Research Forum LLC
Materials Research Foundations**107** (2021) 130-146   https://doi.org/10.21741/9781644901496-6

via carbonation in urban regions [7]. Steel corrosion in concrete could be diminished by implication of numerous recognized strategies [8]; choice of steel that is corrosion-resistant, utilisation of coatings, incorporation of concrete sealers, usage of massive concrete covering, inclusion of corrosion inhibitors or cathodic protection. So as to give extra protection and enhance the life expectancy of reinforced concrete frameworks, various substances recognized as corrosion inhibitors have been employed in current years. Inhibitors can be termed as chemical molecules which eliminates the corrosion and increments the life expectancy of the reinforced concrete arrangement without essentially modifying the amount of added corrosive agent [9]. The utilization of inhibitors is a promising method contrasted to any alternative traditional strategies for safety and rehabilitation, because of their low expense and easy application [10].

Green chemistry has been drawing attention significantly in several areas by scheming chemicals, chemical techniques, and industrial substances with the intention to evade contaminants as well as decrease wastage. One of the branches in which green chemistry is providing development by encouraging the reduction of the ecological effect and the wastes is the field of metallic corrosion. Corrosion of metallic components is significant that is broadly considered an industrial issue, thus it has been discovered as a fruitful analysis area in sustainability. Green inhibitors are drawing considerable focus in the field of corrosion because of their immunity, biodegradability and sustainability [11]. Sustainability in the construction industry could be attained by means of various approaches, with decrease in consumption of gross energy, polluting discharge, and non-renewable natural resources. Considering all these, the following chapter talks about several sustainable corrosion inhibitors established for the concrete, their mechanisms of action and the requirement for green/sustainable corrosion inhibitors for the future with enhanced improvements.

2.   Classification of corrosion inhibitors

Corrosion inhibitor portrays a prominent function in the corrosion phenomenon. Concrete corrosion inhibitors may be distinguished as organic, inorganic and natural inhibitors on the basis of the chemical composition. It is well represented in the schematic flow chart illustrated in Fig.1.

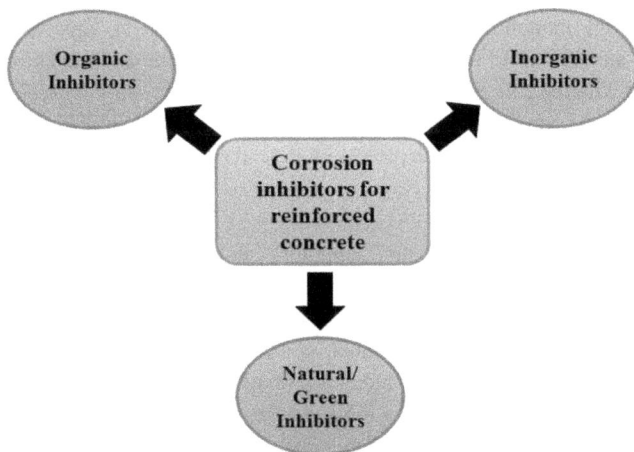

Figure 1.   Schematic flow chart for illustrating the classification of concrete corrosion.

2.1   Organic inhibitors

Amines and alkanolamines have been extensively employed as an organic inhibitors due to their good solubility in water as well as they do not affect the behaviour of concrete [12]. Alkanol amine derived inhibitors might be used to reduce chloride as well as carbonation induced corrosion. Amines and alkanolamines possess unsatisfactory corrosion inhibitory effect as compared to the carboxylates especially polycarboxylates [13]. Carboxylate derived inhibitors will in general reduce the extent of hydration that occurs in concrete, expanding the setting duration [14].

2.2   Inorganic inhibitors

Inorganic inhibitors are incredibly pervasive for providing protection to reinforcement steel corrosion. Calcium nitrite is one of the widely utilised inorganic inhibitor. 4% $Ca(NO_2)_2$ prevents reinforced steel corrosion, that may impart negligible adverse consequence on the quality of the concrete even if the structure is treated in the chloride environment for a longer duration [6]. Although for calcium nitrite derived inhibitors, a prime attention is needed for ascertaining the requisite amount; as if the concentration is lower than the requisite, it would reflect the bad effect together on steel as well as concrete [9]. Many European countries such as Germany and Switzerland have forbidden calcium nitrite as a result of its harmful and lethal reactions [15]. Because of this reason, alternative inhibitors have been considered. Another inorganic inhibitor that has been used is sodium monoflurophosphate (Na_2PO_3F). Na_2PO_3F in aqueous and natural condition is susceptible to hydrolysis, producing orthophosphate in addition to fluoride that interacts with the corrosive substances like Fe_3O_4, c-Fe_2O_3, or $FePO_4.H_2O$. Red mud, residuum from the bauxite is additionally used as an inorganic inhibitor, which possess the ability to improve the corrosion protection in reinforced concrete frameworks [12].

2.3   Natural inhibitors

For the corrosion phenomenon, natural inhibitors can be considered as most effective as well as exceptionally beneficial for the surroundings as compared to other corrosion inhibitors [16]. Researchers nowadays are focusing in the direction of establishing natural strategies for corrosion issues. Their main objective is to develop cost effective, less hazardous, environmental friendly and biocompatible corrosion inhibitors. Few plant extracts have been utilised as natural inhibitors such as polyphenolics, terpenes, alkaloids, flavonoids, and so on which fulfil entire characteristics of natural inhibitors [17]. Sugar portions of vegetable extracts has been tested for providing good inhibition naturally to reinforced steel corrosion [18]. In order to improve the mechanical behaviour of reinforced concrete, Magrabe banana's stem juice has been employed as concrete admixture [19]. To enhance the durability of reinforced concrete deteriorated via sulphate or chloride ions, *Bambusa arundinacea* extract has been used as a green corrosion inhibitor [18].

3.   Mechanism of corrosion in reinforced concrete

For an appropriate comprehension about the mechanisms of the inhibitor's action, it is primarily important to comprehend the process associated with the corrosion of reinforced steel [14]. The corrosion phenomenon of the reinforced concrete framework is an electrochemical procedure that usually happens with the change in the amount of soluble ions in the interior of concrete, building electrochemical potential cells or corrosion cells, outlined with the movement of electrons or ions among the cathodic and anodic zones. Various anodic as well as cathodic reactions are elucidated in equations described below (Eq. 1-5). The schematic representation of the overall reaction mechanism in order to provide the better picture of electrochemical corrosion phenomenon is also represented in Fig. 2.

Anodic reactions

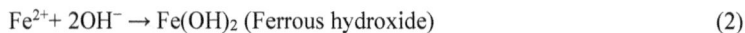

$$Fe \rightarrow Fe^{2+} + 2e^- \tag{1}$$

$$Fe^{2+} + 2OH^- \rightarrow Fe(OH)_2 \text{ (Ferrous hydroxide)} \tag{2}$$

$$4Fe(OH)_2 + 2H_2O + O_2 \rightarrow 4Fe(OH)_3 \text{ (Ferric hydroxide)} \tag{3}$$

$2Fe(OH)_3 \rightarrow Fe_2O_3.H_2O + 2H_2O$ (Hydrated Ferric oxide)   (4)

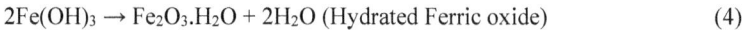

Cathodic reaction

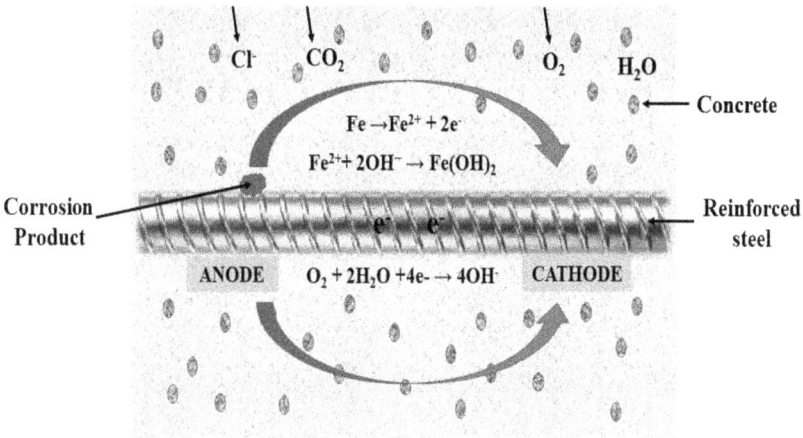

$O_2 + 2H_2O + 4e^- \rightarrow 4OH^-$ (5)

Figure 2.   Model for electrochemical phenomenon for corrosion of steel in concrete.

3.1   Carbonation

Carbonation is a procedure where carbon dioxide gets through the concrete via diffusion reacts with the water existing in the pores, subsequently produces carbonic acid. Carbonic acid further associates to the calcium hydroxide formed at the time of the cement hydration, bringing about the development of calcium carbonate. Such process results in the declination of the surrounding pH to about 8 because the produced $CaCO_3$ is less basic as compared to the calcium hydroxide [20]. Such pH decrease is accountable for the destabilization of the passivation layer, thereby commencing the corrosion phenomenon. The overall carbonation process could be reflected from Eq. 6 and 7:

$CO_2 + H_2O \rightarrow H_2CO_3$ (6)

$H_2CO_3 + Ca(OH)_2 \rightarrow CaCO_3 + 2H_2O$ (7)

3.2   Chloride attack

Chloride attack is primarily the main reason behind the reinforced steel corrosion. The existence of chlorides ions (Cl^-) in concrete usually originates from seawater, coastal areas, setting time accelerator additives having $CaCl_2$, water or pollutants from industrial contaminants and they may be attributed as the main causes of reinforced concrete corrosion [21]. Chloride ions penetrates to the concrete through the pore network and micro cracks, producing the oxide film over the reinforcing steel and therefore speeds up the reaction of corrosion and concrete deterioration [22]. Appearance of chloride ions in the solution encompassing iron, which binds to Fe^{2+} of passive layer onto the metallic surface and develops an iron–chloride complex (Eq. 8). Further hydrolysis of such complex leads to ferrous hydroxide in addition to chloride ions that are available for additional attack above the steel surface (Eq. 9).

$$Fe^{2+} + Cl^- \rightarrow [FeCl \text{ complex}]^+ \qquad\qquad (8)$$

$$[FeCl \text{ complex}]^+ + 2OH^- \rightarrow Fe(OH)_2 + Cl^- \qquad\qquad (9)$$

4.   Recent developments on sustainable corrosion inhibitors for concrete

Usually, corrosion inhibitors for reinforced steel in concrete are incorporated single time in the structure and an efficient inhibitor for reinforced concrete must possess following characteristics:

- efficient even at very less concentrations
- long term endurance in the interior of the solid
- homogeneous distribution
- They may be easily detachable from concrete and should not affect the characteristics of the concrete.

Ecological issues are expanding globally and may probably going to impact the selection criteria for corrosion inhibitors in the next generation. Eco-friendly and cost effective inhibitors which may overcome the requirement for synthesis [23] were demonstrated to be productive and exceptionally advantageous for the surroundings and are proven to be sustainable contrasted to few organic as well as inorganic inhibitors [23]. Plant sources fulfil all basic necessities as their derivatives specifically alkaloids, flavonoids, terpenes or polyphenolics are highly enriched with electron donating hetero groups (S, N, O and conjugated π electrons). These plants are rich source of naturally produced chemical substances, which may be removed via easier and low-cost processes, besides being biodegradable. Therefore, plant sources can be utilized to produce natural and sustainable corrosion inhibitors, hence its current advancements has been investigated in several papers [24,25]. In view of this, biomass or renewable raw substances can be employed for several industrial applications to promote green chemistry and sustainability. One of the areas where green chemistry, usually related to biomass-derived substance, is establishing advancements, by decreasing the ecological impact and the wastes, is linked to the safeguarding metals. Fig. 3 illustrates the use of natural sources and biomass wastes extracts as sustainable corrosion inhibitors for steel reinforced corrosion. The significance of developing bio-wastes is important in the field of corrosion. Few works that studied the extracts from natural residues or bio-wastes, that are usually scrapped in the surroundings otherwise, are employed as corrosion inhibitors by many researchers.

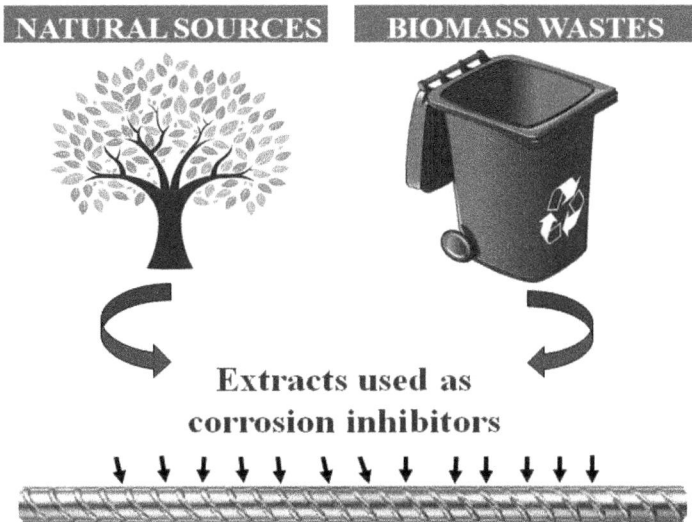

Figure 3.   *Extracts used as sustainable corrosion inhibitors for steel reinforced corrosion.*

Quraishi et al. [26] explored the effects of calcium palmitate solely and in aggregation with calcium nitrite on the corrosion of steel in concrete. The results of the study proved calcium palmitate as a proficient and promising inhibitor; providing 91 to 92% efficiency after 90 days of treatment in 3.5% NaCl environment. It was also shown that the inhibitor do not have any effect on the mechanical durability of cement as well as concrete. Petro graphic study revealed that calcium palmitate blocks the pores, subsequently reducing the extent of corrosion. Further investigation suggested that calcium palmitate inhibited corrosion through adsorption mechanism. Inhibitor created the film over the steel surface by means of polar carboxylate group obstructing the pores and thus producing insoluble hydrophobic ferric stearate salt.

Palanisamy et al. [27] analysed the activity of another plant extract i.e. Prosopis juliflora on the corrosion of reinforced steel in 3.5 % NaCl. Corrosion tests were conducted for 30 days immersion time of rebar without and with the inhibitor. From corrosion tests, it was revealed that the corrosion inhibitor reached the efficiency of 91% at 120 ppm concentration of inhibitor.

Zhang and his co-authors [28] examined the corrosion inhibitory behaviour of maize gluten meal extracts for steel in simulated concrete pore solution having 3.0 wt% NaCl. The outcomes of EIS and polarization measurements have been found to be in good correlation with each other that suggested an increment in the corrosion inhibition of steel after addition of extract relatively to the solution. Adsorption of inhibitor on the metallic surface obeyed Langmuir adsorption isotherm which may be primarily due to physisorption. Moreover, the amide moieties in the primary components of extract were additionally helpful to provide adherence to the surface of steel. The sustainable extract is obtained as a by-product from starch manufacturer that is proven to be an efficient and natural inhibitor for combating concrete corrosion having efficiency of 62.71–88.10%.

Pradipta and co-workers [29] showed that at same volume, green tea extract (GT) displayed an elevated inhibition efficiency (I.E.) as compared to widely used calcium nitrite inhibitor for the corrosion of reinforced steel inserted in mortar. GT acted as a mixed type corrosion inhibitor that enhanced rebar polarization resistance (R_p); reflecting the generation of a protective covering on steel surface. After the development of such layer, rebar corrosion was diminished. Surface studies indicated that the film was rich in calcium, particularly the calcium carbonate polymorphs.

Anitha et al. [30] assessed Rosa damascena leaves as a green corrosion inhibitor for reinforced steel corrosion (Fig. 4). The anticorrosion performance has been estimated using electrochemical impedance technique in addition to polarisation measurements in simulated pore solution. The adsorption phenomenon followed Langmuir adsorption isotherm indicating the physisorption mechanism. The highest inhibition efficiency was found to be 82% having 12 v/v % compositions. SEM representations described the chelating behaviour of phytochemical components comprising hetero atoms like Oxygen, Nitrogen or aromatic and pi electrons associating to iron ions and thereby promoting the protection for metallic corrosion.

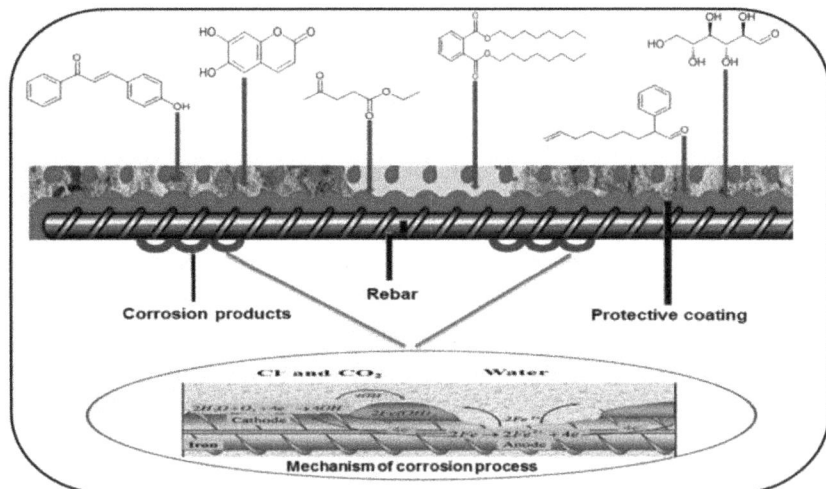

Figure 4.　Illustration of mechanism of Rosa damascene as sustainable/green corrosion inhibitor for reinforced steel corrosion. Reproduced with permission from [30]; Copyright 2019 @ Elsevier.

Replacement of cement powder with sustainable waste products is the primary attention of research in concrete field. Loto and his co-authors [31] investigated cow bone ash (CBA) for estimating anticorrosion behaviour on mild steel corrosion in artificial concrete pore solution using potentiodynamic polarization study and current–time displacement studies. Recorded data reflected that CBA extensively diminished the corrosion process with highest inhibition efficiency of 80.18% having 20% CBA showed anodic sort of inhibition. Also, the cathodic side of the polarization plot depicted eminent reduction in slope after 0% CBA.

Shubina et al. [32] reported biomolecules as potential environmental friendly inhibitor for steel reinforcement in concrete. Biomolecules employed in present investigation are one of the new categories of substances generated from bacterial cells. The inhibition behavior in simulated concrete pore solution has been illustrated utilizing classical electrochemical techniques and microscopic studies. The reduction in corrosion current density was found to be significant after including 1 gL^{-1} of inhibitor in the simulated concrete pore solution. Linear Polarization measurements as well as Electrochemical Impedance analysis reflected comparatively better inhibition performance ($\approx$58.6%).

Furthermore, XPS studies and SEM results depicted that such kind of biomolecules might be considered as film producing mixed type inhibitor.

Abbas et al. [33] investigated the performance of green inhibitor extracted from the waste of orange peels. They extracted dry orange peel using methanol extract at 6-hour immersion time in methanol with the pressure of 60 mbar and 40°C. From the experimental data like electrochemical polarisation measurements and weight-loss testing performed for 7 days immersion time of rebar, it was demonstrated that this inhibition showed a good performance in aqueous 3.5 wt% NaCl solutions. The steel rebar showed that the corrosion rates of rebar reduced to 0.02 mm/year at 3% composition of inhibitor.

Akshatha et al [34] performed studies on inhibitor with leaves of ruta graveolens and azadirachta. The extracted leaf of Azadirachta indica (neem) and ruta graveolens plants were used as organic inhibitors. Comparison with inorganic inhibitor like sodium nitrate and ethylene diamine tetra acetic (EDTA) disodium dehydrate was also done. The inhibitors were added during mixing of concrete and the reinforcing steel bar. The concrete was made of cement of OPC 43 grade with specific gravity 3.279. The rebar was dipped in hydrochloric acid (HCl), sodium chloride (NaCl) and magnesium sulphate (MgSO$_4$) solution for performing corrosion tests. From half-cell potential measurement, experiment in 5% HCl for 56 days, it can be seen that extracts of Azadirachta Indica have shown the most positive potential, followed by ruta graveolens, sodium nitrate and EDTA disodium dihydrate. The same trend was also observed when the experiments were conducted in 5% NaCl solutions which illustrated that Azadirachta Indica gave the most positive corrosion potential.

Shaymaa et al. [35] studied powdered rice husk for corrosion of steel in concrete. Extract of rice husk was included to the concrete with American mix design method (ACI 211) with the strength of 30 MPa in 28 days. The samples of rebar were treated with tap water for 30 days and immersed in 3.5% NaCl environment. Corrosion experiments were conducted for the solution of varied concentrations (1%, 2% and 3%) of inhibitor. The corrosion current was determined to be 41.3 μA/cm^2 for the solution without inhibitor which becomes 28.5 μA/cm^2 and 7.8 μA/cm^2 for the solution having 1% and 3% rice husk powder concentration, respectively. It signified that the reduction of corrosion rate was 30% and 81% respectively

Joshua et al. [36] characterized phyllanthus muellerianus as an inhibitor that promote sustainability for combating concrete steel-reinforcement corrosion in industrial condition. They used 0.5M H$_2$SO$_4$ media to simulate industrial/microbial condition. At the concentration of 6.67 g/L, this inhibitor reduced corrosion rate of rebar to 90%. While at the concentration of 1.67%, the reducing corrosion rate was found out to be 78%. From investigation, leaves of phyllanthus muellerianus and euphorbiaceae contained tannins, phlobatanins, saponins, flavooids, terpenoids and alkanoids as major constituents.

Vernonia amygdalina extract has been examined as green corrosion inhibitor for mild steel rebar concrete in 3.5M NaCl solution by Loto et al. [37]. Corrosion resistance of Vernonia amygdalina was analysed using weight loss technique, potential & pH estimations and several concrete parameters were further discussed. The inclusion of Vernonia amygdalina extract into the solution significantly changed the potential, pH and compressive strength.

Okeniyi et al. [38] investigated Rhizophora mangle L extract which behaved as effective corrosion inhibitor admixture for reinforcing steel in 0.5M H_2SO_4. The electrochemical outcomes were analysed by using statistical distribution fitting models and subsequent inhibition efficiencies were further estimated. Inhibition efficiencies outcomes by experimental procedure and theoretical calculations has been plotted with respect to various amounts of extract; correlation and conflict among experimental data and theoretical assumed data has been also evaluated.

Eyu et al. [39] discussed the anticorrosion performance of Vernonia amygdalina extract for carbon steel in concrete in 3.5M NaCl solution. Corrosion study was performed using weight loss experiments, corrosion potential calculations, half-cell calculations, concrete resistivity measurement as well as visual analysis. The outcomes acquired were also correlated with commercially available inhibitors like sodium nitrite or calcium nitrites. Phytochemical constituents of Vernonia amygdalina extract were described to contain alkaloids, saponins and tannins. Corrosion tests may be illustrated by the plots against weight loss, inhibition efficiency as well as corrosion rate. Outcomes achieved evidently discussed that Vernonia amygdalina extract showed analogous corrosion resistance when compared to commercially available inhibitors (Sodium nitrate and Calcium Nitrite).

Abdulrahman et al. [40] has broadly discussed and illustrated the anticorrosion behaviour of Bambusa arundinacea (BA) extract for reinforced steel corrosion in concrete. It was investigated to enhance the strength of concrete polluted by chloride or sulphate environment. The obtained data was correlated with known Calcium nitrite and ethanolamine inhibitors. Effect of inclusion of $MgCl_2$ and $MgSO_4$ over concrete strength had been estimated and elucidated. Fundamental advantages of present analyses illustrate that inclusion of inhibitor shows negligible adverse effect on concrete durability. BA did not comprise any heavy metal that also eliminates the risk of production of dioxin on interacting with chloride contaminated water.

Lisha et al. [41] used azadirachta indica (neem) powder and dehydrated aloe vera as corrosion inhibitor for steel in concrete. M 25 grade concrete was used with coarse aggregates of 20 mm size. The concrete was immersed in solution with salinity of about 3.5% (35 g/L). The concrete with green corrosion inhibitors has reduced corrosion rate of steel on concrete from 0.3 to 0.22 mm/y. Aloe vera inhibitor extract was able to diminish corrosion rate to 0.27 mm/y. The results of study showed that azadirachta indica (neem) reflects better corrosion inhibition efficiency when compared to the aloe vera inhibitor.

Conclusions and outlook

Usually, inorganic/chemical inhibitors have been employed in corrosion protection of steels in concretes. However, due to its hazardous nature for the environment, usages of such inhibitors are going to be confined. As a substitute, green/sustainable corrosion inhibitors have gained much interest in choosing inhibitors for reinforced steel. Such green inhibitors fulfil all the needed features of inhibitors employed for steel in concretes that includes lower cost, compatibility with concretes, non-toxicity, biodegradability and applicability for industrial constructions. Mechanism theories of inhibitors in combating corrosion phenomenon in steel rebar include formation of barrier film over the metallic surface via physical/chemical reactions among steel and concretes. Therefore, they don't allow the penetration of corrosive ions to the metallic surface by obstructing concrete pores. Additionally, inhibitors could interfere anodic or cathodic reactions which diminish the corrosion rates. Subsequently, the demand of eco-friendly inhibitors for steel rebar has enhanced significantly. From the literature, several evidences conclude that the plant or biomass waste extracts might be used as excellent corrosion inhibitors for steel reinforced corrosion. Moreover, competing with the higher inhibition efficiency obtained by ideal inorganic/chemical inhibitors should be considered in the future and may be a significant area of future research. But, from the studies, it can be confidently concluded that green inhibitors will continue to provide corrosion protection in the future with further modifications. Future improvements may include development of modified corrosion inhibitors possessing excellent adsorption ability to attain more attractive anticorrosion performance.

Acknowledgement

All authors are thankful to their institutes for providing facilities to carry out this work.

References

[1]   M.G. Fontana, Corrosion Engineering, third ed., McGraw-Hill Series in Materials Science and Engineering, 1987.

[2]   C.L. Page, K.W.J. Treadaway, Aspects of the electrochemistry of steel in concrete, Nature. 297 (1982). https://doi.org/10.1038/297109a0

[3]   V. Elfmarkova, P. Spiesz, H.J.H. Brouwers, Determination of the chloride diffusion coefficient in blended cement mortars, Cem. Concr. Res. 78 (2015) 190–199. https://doi.org/10.1016/j.cemconres.2015.06.014

[4]   A.S. Abdulrahman, M. Ismail, M.S. Hussain, Corrosion inhibitors for steel reinforcement in concrete : A review, Sci. Res. Essays. 6 (2011) 4152–4162. https://doi.org/10.5897/SRE11.1051

[5]   F. Fei, J. Hu, J. Wei, Q. Yu, Z. Chen, Corrosion performance of steel reinforcement in simulated concrete pore solutions in the presence of imidazoline quaternary ammonium salt corrosion inhibitor, Constr. Build. Mater. 70 (2014) 43–53. https://doi.org/10.1016/j.conbuildmat.2014.07.082

[6]   H.-S. Lee, H.-S. Ryu, W.-J. Park, M.A. Ismail, Comparative study on corrosion protection of reinforcing steel by using amino alcohol and lithium nitrite inhibitors, Materials (Basel). 8 (2015) 251–269. https://doi.org/10.3390/ma8010251

[7]   E. Redaelli, L. Bertolini, Electrochemical repair techniques in carbonated concrete . Part II : cathodic protection, J. Appl. Electrochem. 41 (2011) 829–837. https://doi.org/10.1007/s10800-011-0302-3

[8]   V.S. Sastri, E. Ghali, M. Elboujdaini, Corrosion Prevention and Protection — Practical Solutions, John Wiley Publications, Great Britain, 2007. https://doi.org/10.1002/9780470024546

[9]   C. Alonso, M. Sanchez, C. Andrade, J. Full, Protection capacity of corrosion inhibitors against the corrosion of rebars embedded in concrete, in: J.M. Costa (Ed.), Trends Electrochem. Corros. Begin. 21st Century Ed., I Edicions De La Universitat De Barcelona Publications, Barcelona, 2004: p. 586.

[10]   T.A. Soylev, M.G. Richardson, Corrosion inhibitors for steel in concrete : State-of-the-art report, Constr. Build. Mater. 22 (2008) 609–622. https://doi.org/10.1016/j.conbuildmat.2006.10.013

[11]   D. Darling, R. Rakshpal, Green chemistry applied to corrosion and scale inhibitors, in: Corrosion, NACE International, Houston, TX, US, 1998.

[12]   M.L.S. Rivetti, J. da S.A. Neto, N.S. de A. Júnior, D.V. Ribeiro, Corrosion Inhibitors for Reinforced Concrete, in: Corros. Inhib. Princ. Recent Appl., 2018. https://doi.org/10.5772/intechopen.72772

[13]   B. Elsener, U. Angst, Corrosion inhibitors for reinforced concrete, in: Sci. Technol. Concr. Admixtures, Elsevier Ltd, 2015. https://doi.org/10.1016/B978-0-08-100693-1.00014-X

[14]   R. Myrdal, Corrosion Inhibitors – State of the art, Oslo: SINTEF Building and Infrastructure. COIN Project Report; 22 (2010).

[15]   P.B. Raja, M.G. Sethuraman, Inhibitive effect of black pepper extract on the sulphuric acid corrosion of mild steel, 62 (2008) 2977–2979. https://doi.org/10.1016/j.matlet.2008.01.087

[16]   A.A. Salawu, M. Ismail, M. Zaimi, A. Majid, Z.A. Majid, C. Abdullah, J. Mirza, Green bambusa arundinacea leaves extract as a sustainable corrosion inhibitorin steel reinforced concrete, J. Clean. Prod. 67 (2014) 139–146. https://doi.org/10.1016/j.jclepro.2013.12.033

[17]   D. Kesavan, M. Gopiraman, N. Sulochana, Green inhibitors for corrosion of metals : A Review, Chem. Sci. Rev. Lett. 1 (2012) 1–8.

[18]   P.B. Raja, S. Ghoreishiamiri, Natural corrosion inhibitors for steel reinforcement in concrete- A Review, Surf. Rev. Lett. 22 (2015) 1550040-1550048. https://doi.org/10.1142/S0218625X15500407

[19]   N.S. Berke, A. Rosenberg, Technical review of calcium nitrite corrosion inhibitor in concrete, Transp. Res. Rec. 1211 (1989) 18–27.

[20]   Z. Shi, B. Lothenbach, M. Rica, J. Kaufmann, A. Leemann, S. Ferreiro, J. Skibsted, Experimental studies and thermodynamic modeling of the carbonation of Portland cement , metakaolin and limestone mortars, Cem. Concr. Res. 88 (2016) 60–72. https://doi.org/10.1016/j.cemconres.2016.06.006

[21]   M. Ormellese, M. Berra, F. Bolzoni, T. Pastore, Corrosion inhibitors for chlorides induced corrosion in reinforced concrete structures, Cem. Concr. Res. 36 (2006) 536–547. https://doi.org/10.1016/j.cemconres.2005.11.007

[22]   L. Chung, J.J. Kim, S. Yi, Bond strength prediction for reinforced concrete members with highly corroded reinforcing bars, Cem. Concr. Compos. 30 (2008) 603–611. https://doi.org/10.1016/j.cemconcomp.2008.03.006

[23]   J. Olusegun, O. Cleophas, A. Loto, A. Patricia, I. Popoola, Electrochemical performance of anthocleista djalonensis on steel- reinforcement corrosion in concrete immersed in saline / marine simulating-environment, Trans. Indian Inst. Met. 67 (2014) 959–969. https://doi.org/10.1007/s12666-014-0424-5

[24]   D. Kesavan, M. Gopiraman, N. Sulochana, Green inhibitors for corrosion of metals : A Review, Chem. Sci. Rev. Lett. 1 (2012) 1–8.

[25]   J. Buchweishaija, Phytochemicals as green corrosion inhibitors in various corrosive media: A Review, Tanzania J. Sci. 35 (2008) 77–92.

[26]   M.A. Quraishi, V. Kumar, B.N. Singh, S.K. Singh, Calcium Palmitate : A green corrosion inhibitor for steel in concrete environment, J. Mater. Environ. Sci. X (2012) 1001–1008.

[27]   S.P. Palanisamy, G. Maheswaran, C. Kamal, G. Venkatesh, Prosopis juliflora — A green corrosion inhibitor for reinforced steel in concrete, Res. Chem. Intermed. 42 (2016) 7823–7840. https://doi.org/10.1007/s11164-016-2564-1

[28]   Z. Zhang, H. Ba, Z. Wu, Sustainable corrosion inhibitor for steel in simulated concrete pore solution by maize gluten meal extract : Electrochemical and adsorption behavior studies, Constr. Build. Mater. 227 (2019) 117080. https://doi.org/10.1016/j.conbuildmat.2019.117080

[29]   I. Pradipta, D. Kong, J. Ban, L. Tan, Natural organic antioxidants from green tea form a protective layer to inhibit corrosion of steel reinforcing bars embedded in mortar, Constr. Build. Mater. 221 (2019) 351–362. https://doi.org/10.1016/j.conbuildmat.2019.06.006

[30]   R. Anitha, S. Chitra, V. Hemapriya, I. Chung, S. Kim, M. Prabakaran, Implications of eco-addition inhibitor to mitigate corrosion in reinforced steel embedded in concrete, Constr. Build. Mater. 213 (2019) 246–256. https://doi.org/10.1016/j.conbuildmat.2019.04.046

[31]   R.T. Loto, A. Busari, Electrochemical study of the inhibition effect of cow bone ash on the corrosion resistance of mild steel in artificial concrete pore solution, Cogent Eng. 6 (2019) 1644710. https://doi.org/10.1080/23311916.2019.1644710

[32]   V. Shubina, L. Gaillet, T. Chaussadent, T. Meylheuc, J. Creus, Biomolecules as a sustainable protection against corrosion of reinforced carbon steel in concrete, J. Clean. Prod. 112 (2015) 666–671. https://doi.org/10.1016/j.jclepro.2015.07.124

[33]   A.S. Abbas, É. Fazakas, T.I. Török, Corrosion studies of steel rebar samples in neutral sodium chloride solution also in the presence of a bio-based (green) inhibitor, Int. J. Corros. Scale Inhib. 7 (2018) 38–47.

[34]   G.M. Akshatha, B.G.J. Kumar, H. Pushpa, Effect of corrosion inhibitors in reinforced concrete, Int. J. Innov. Res. Sci. Eng. Technol. 4 (2015) 6794–6801. https://doi.org/10.15680/IJIRSET.2015.0408013

[35] S.A. Abdulsada, A.I. Al-mosawi, A.A.A. Hadi, Studying the effect of eco-addition inhibitors on corrosion resistance of reinforced concrete, Bioprocess Eng. 1 (2017) 81–86.

[36] J.O. Okeniyi, O.A. Omotosho, O.O. Ogunlana, E.T. Okeniyi, T.F. Owoeye, A.S. Ogbiye, E.O. Ogunlana, Investigating prospects of phyllanthus muellerianus as eco-friendly / sustainable material for reducing concrete steel- reinforcement corrosion in industrial / microbial environment, Energy Procedia. 74 (2015) 1274–1281. https://doi.org/10.1016/j.egypro.2015.07.772

[37] C.A. Loto, O.O. Joseph, R.T. Loto, A.P.I. Popoola, Inhibition effect of vernonia amygdalina extract on the corrosion of mild steel reinforcement in concrete in 3.5m nacl environment, Int. J. Electrochem. Sci. 8 (2013) 11087–11100.

[38] J.O. Okeniyi, C.A. Loto, A.P.I. Popopla, Corrosion inhibition performance of rhizophora mangle l bark-extract on concrete steel-reinforcement in industrial / microbial simulating-environment, Int. J. Electrochem. Sci. 9 (2014) 4205–4216.

[39] D.G. Eyu, H. Esah, C. Chukwuekezie, J. Idris, I. Mohammad, Effect of green inhibitor on the corrosion behavior of reinforced carbon steel in concrete, ARPN J. Eng. Appl. Sci. 8 (2013) 326–332.

[40] A.S. Abdulrahman, Green plant extract as a passivation- promoting inhibitor for reinforced concrete, Int. J. Eng. Sci. Technol. 3 (2011) 6484–6490.

[41] C. Lisha, M.Rajalingam, S. George, Corrosion resistant of reinforced concrete with green corrosion inhibitors, Int. J. Eng. Sci. Invent. Res. Dev. 3 (2017) 687–691.

Sustainable Corrosion Inhibitors   Materials Research Forum LLC
Materials Research Foundations107 (2021) 147-174   https://doi.org/10.21741/9781644901496-7

Chapter 7

Green Corrosion Inhibitors for Coatings

R. Payal[1], A. Jain[2]*

[1]Rajdhani College, Department of Chemistry, University of Delhi, Delhi 110007

[2]Daulat Ram College, Department of Chemistry, University of Delhi, Delhi 110007

* jainarti21@gmail.com

Abstract

Corrosion is emerging as a potential hazard which abolishes metals and their structures and hence become an imperative menace. It is an omnipotent and omnipresent process which is present in every environment, *i.e.*, air, soil, water. Green chemistry is one of the notable branches of chemistry that focuses on the protection of environment and human well–being in an economically viable approach allowing dodging of toxins and reducing hazards due to corrosion. Green chemistry exploited well–known strategy namely green inhibitors to prevent, control or impede the growth of corrosion. Green inhibitors are eco–friendly, cost–effective, renewable natural products which are favourable over toxic synthetic corrosion inhibitors. Extracts of natural products contain natural products containing alkaloids, carboxylic acids, nicotine, polyphenols, quinine, terpenes, and other functional groups possessing elements like C, N, O, S, *etc.*, prompting adsorption *via* forming a thin layer (coating) on the metallic surface to shield the surface and encumber corrosion. In the field of economical loop, this approach develops various potential applications in manufacturing areas other than 'Trash to treasure'. Even though a bunch of experiments have been performed and several research articles have been in print, however, the area of green inhibitors is still demanding more investigation on this open issue. More and more interest in the area extended the research, consequentially to a large variety of tried molecules. Nevertheless, the most accepted protocols are classical and, therefore, are incompetent to completely portray the probable worth of inhibitors. Hence all above stated features should be the objective of the contemporary research so that productive analysis to emphasize the weak areas of the green inhibitors field and tackle the prospect research in the field that still requires validation.

Keywords

Anodic Inhibitor, Cathodic Inhibitor, Green Corrosion Inhibitors, Coatings, Ionic Liquids, Vapor–Phase Inhibitor, Sustainable

Sustainable Corrosion Inhibitors   Materials Research Forum LLC
Materials Research Foundations**107** (2021) 147-174   https://doi.org/10.21741/9781644901496-7

Contents

1.   Introduction

Corrosion is derived from Latin word corrōdere which means "to gnaw". Corrosion is a universal phenomenon where living tissues, metals, alloys, ceramic, polymer or materials are consumed by virtue of chemical or electrochemical reaction. In this process materials attempt to return to their most stable thermodynamics state as a result of their reaction with the environment medium [1]. The more corrosive environment comprises of air, carbon dioxide, gaseous sulphur, molten salts, organic liquids, water, whereas, innocuous environments include gamma radiation, nuclear fission fragments, neutron beams, and ultraviolet light. Materials such as alloys, ionic and covalent solids, rubber, plastics, metals, composite materials concrete and wood are highly subjected to corrosion [2]. Corrosion is an exorbitant growth as a consequence of damaging resources or their properties (estimated economy expenditure of various countries due to corrosion is mentioned in Table 1), which in turn, causes shut down or severe debacle of systems, excessive time loss throughout maintenance. In some cases, injuries and several hazardous effects on living systems have been reported [3-7].

Table 1.   *Corrosion cost in different countries.*

Country	Cost (in $)	Reference
Australia	470 million	[8]
Finland	47–62 million	[9]
India	320 million	[10]
Japan	9.2 billion	[11]
Russia	6.7 billion	[12]
Sweden	58–77 million	[13]
United Kingdom	3.2 billion	[14]
Unites state of America	276 billion	[15]
West Germany	1.5 billion	[16]

Different kind of corrosion, their carriers and various synthetic organic and inorganic corrosion inhibitors are mentioned in literature (Fig. 1, Table 2) [17–20], however, hazardous effects triggered by these toxic regular inhibitors paved way for green corrosion inhibitors derived from eco-friendly and biodegradable natural products. Plant extracts are widely used as green corrosion inhibitors as they reckoned to be bounteous in producing chemical compounds [21–23]. Generally, small amount of corrosion inhibitors are added to the corrosive surface of alloys or metals to decrease the rate of corrosion *via* accumulation of operative species onto affected metal surfaces by: 1) altering the anodic or cathodic cell reactions, 2) trifling the interaction between violent ions and metal surfaces, 3) enhancing metallic surface's electrical resistance by means of creating a protective layer (coating) onto the surface [2].

In this chapter different categories of corrosion, and various sources of green corrosion inhibitors has been discussed. Different natural products which are utilized for preparation of corrosion inhibitors are mentioned. Here, we will be focussing on the coating of metallic surface to prevent them from erosion. The mechanism outline has also included.

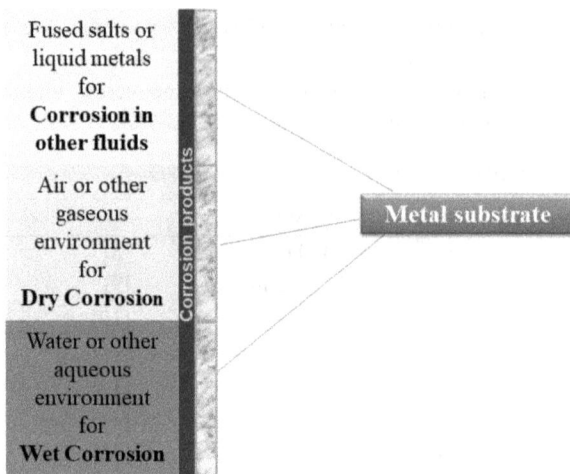

Figure 1.   Corrosion under different environment conditions [24].

2.   Classifications of inhibitors

Corrosion inhibitors are chemical reagents that reduce the rate of metal depletion on adding to a corrosive aqueous environment in smaller quantity. The inhibitors are classified into different categories depending on their mode of action:

2.1   Anodic inhibitor

Anodic inhibitors are liable to hinder the anodic reaction by forming either a deposit layer or a passive layer on the anode by the deposition of sparingly soluble species, like oxides, hydroxides, or salts. These are operative in the pH range of ~6.5–10.5. Examples of anodic inhibitors include NO_2^-, MoO_4^{2-}, CrO_4^{2-}, borates, molybdates, phosphates, tungstates, etc. however, they suffer a serious drawback in terms of accelerating corrosion due to poor management and hence, not fit for oil and gas industry. Inorganic anodic inhibitors offer approximately 80–95% prevention.

2.2   Cathodic inhibitor

Cathodic corrosion inhibitors inhibit cathodic reaction either by selective precipitation on the cathodic side to decrease the dispersion of deoxidized species or by decelerating their reactions by creating a shielding on cathodic zones for counterattack on hydrogen and oxygen in acidic–and alkaline conditions, respectively. Examples of Cathodic inhibitors are oxygen scavengers, cathodic poisons, cathodic precipitates, borates or silicates in alkaline solutions,

Table 2.   Different Types of Corrosion [25-29]

Corrosion	Type	Description
Localized	Crevice	Concentration cell corrosion owing to the trapping of corrosive liquid within the voids of the metal.
	Pitting	Voids are formed due to attacks on the selective potions of metal surface. The voids and unharmed fragment of the metal worked as anode and cathode, respectively.
	Filiform	Cell corrosion on surface of metal covered with a thin layer of organic film.
General	Galvanic	Result from interaction of electrolytes with different metals that varies in electrical potentials.
	Uniform	Deteriorates the whole surface or large fraction of the metal and makes the surface thin.
Microbiological	–	Loss of metal surface initiated by biological organisms and can ensue in any aqueous environments owing to omnipresence of microbes and adequate nutrients in fluids.
Metallurgical influenced	Dealloying	Capable of corroding various components of an alloy.
	Exfoliation	Occurs when corrosion propagates along intergranular paths parallel to the material surface.
	Intergranual	Corrosion ensues on the edges or the grain boundaries of the surface of metal.
Q1	Erosion	Movement supported corrosion owing to flow of corrosive liquids on the surface of metal.
	Corrosion fatigue	A combination of cyclic stress and corrosion.
	Wear	Wear Corrosion is defined as the damage caused by synergistic attack of surface failure resulting from dynamic reaction between surface and corrosion in a corrosive environment.

Environmental induced	Stress corrosion cracking	A complex type of corrosion as a result of corrosive environment and pressure.
	Hydrogen Embrittlement	Heat treatment to remove absorbed hydrogen.
	Hot cracking	Formation of shrinkage cracks during the solidification of weld metal.
	Liquid metal Embrittlement	Cracking caused by contact with a liquid metal.

inorganic phosphates, polyphosphates ($Na_4(P_2O_7)_n$), *etc.* Cathodic inhibitors function by obstructing hydrogen growth in acidic medium or by reducing the oxygen concentration in alkaline or neutral solutions. Rare earth salts also find mention as cathodic inhibitor which forms a thin film of its metal oxide at cathodic sites as a result of enflamed alkalinity at these sites due to oxygen diminished reaction and increased solubility product of these metal oxide [2].

2.3   Adsorptive corrosion inhibitors or mixed inhibitors

Additives that cause stimulation of both anodic and cathodic reactions are known as mixed inhibitors. These inhibitors adsorb on anode and cathode (metal surface) by chemisorption or physisorption and block the contact with the metal. The organic molecules having functional groups such as phosphates ($-PO_3{}^{2-}$), carboxylic acids ($-COOH$), amino ($-NH_2$) as well as multiple bonds in their structure form chemisorbed bonds with metal surfaces [30].

2.4   Barrier layer forming inhibitors

Corrosion inhibitors are further categorized into oxidizers, conversion layer formers and adsorbed layer former. These inhibitors form a molecular barrier on the surface of metal and are efficient in decreasing both cathodic and anodic reaction rates except the oxidizers, which transfer the corrosion tendency of the metal to specified site at which a steady oxide or hydroxide is formed and save the metal surface [31]. Santos and coworkers described displacing adsorbed water molecules from the metallic surface with soluble organic compounds inhibitors [32].

$$Org_{(sol)} + nH_2O_{ads} \quad \rightarrow \quad Org_{ads} + nH_2O_{(sol)}$$

2.5   Neutralizing inhibitors

These types of inhibitors take out the hydrogen ions (H^+) from the corrosive atmosphere and hence decreasing the tendency of corrosion. These inhibitors are generally employed for the treatment of boiler waters and in oil field work.

2.6   Scavengers

These are used to eliminate corrosive species other than hydrogen ions (H^+). General example is hydrazine (NH_2NH_2) in boiler systems, which eliminates traces of cathodic reactant *viz.*, oxygen.

Other than these classifications, corrosion inhibitors can also be categorized as inorganic and organic inhibitors attributed to hard and soft acid and base theory (HSAB), and are labelled as hard, soft, and borderline inhibitors. Approximately 80% of organic inhibitors come under this category.

3.   Selection of corrosion inhibitors

The phenomenon of adsorption forms the basis of the corrosion inhibition in which inhibitor gets accumulated onto the metal surface, thus, shielding the surface from encountering corrosion or by retarding electrochemical processes [33]. The corrosion inhibitors can be chosen on the basis on 1) structural features, and 2) theoretical considerations.

3.1   Structural features

In 1954 Hackerman proposed the idea of a 2–D virtual complex based on strength and stability of metal–inhibitor adsorbate complex [34]. The surface coating hinders either the anodic reaction (known as anodic inhibitors) or the cathodic reaction (alias cathodic inhibitors), or both. The coating on the surface also blocks the transference of reactant onto the product or stimuluses the rate constant, which results into corrosion inhibition. One of the publications established *ortho*–substituted organic compounds, *for ex*ample, *ortho*–toluidine as more effective corrosion inhibitor comparative to *meta*– and *para*– substituted toluidines in oil reservoirs [35]. The outcomes of structural variation and the impact of C–N–C bond angles on the backbone of organic compounds on corrosion inhibition was evaluated by altering the methylene rings of polyethylene imines. These factors strike the accessibility of the electron pairs and therefore, change the magnitude of corrosion inhibition efficacy.

Sastri [2] also revealed that a mixture of cerium chloride in 0.1M sodium chloride solution diminish rate of deterioration of Al–Zn–Mg–Cu composite by ~2 times. Popova *et al.* [36] examined some benzimidazoles as corrosion inhibitor for mild steel in deaerated 1 mol L^{-1} HCl solution. They testified that these benzimidazoles have noticeable corrosion inhibition properties. Subsequently, Popova and teammates [37] also showed 8–benzimidazole derivatives as pronounced mixed corrosion inhibitor for mild steel in acidic medium at 20°C. Their effectiveness increases with increased concentration of the additive. Furthermore, Popova *et. al.* [38] also recognized azoles as mixed corrosion inhibitor for mild steel. Samardzija and associates [39] investigated four alcohols as corrosion inhibitors for iron in 1 mol L^{-1} HCl at ambient temperatures and found that these reagents worked as mixed inhibitors and their efficiency differs on the chain length as well as location of the triple bond. Finšgar *et. al.* [40] discovered a corrosion inhibitor formulation, a blended mixture of solvent, surfactant, and an intensifier with the aim of increasing the effectiveness of a corrosion inhibitor in protecting steel. The residuary charge around the metal as well as inhibitor's chemical structure facilitates the surface assimilation of inhibitors on the surface of metal.

In major applications, environment friendly, inhibitors, such as tungstates, and molybdates are widely used which have replaced governmental agencies banned organic inhibitors like chromates. Some common inhibitors such as silicates, carbonates, and borates, are still in use. Inorganic inhibitor borates provide corrosion defence by enhancing the pH of the system [2]. Therefore, it can be established that geometry, steric effects, stoichiometry and backbone of a cluster played a crucial role in designing the efficacy of anti–corrosion compounds.

3.2   Theoretical considerations

Computational tools (Quantum chemical and Molecular dynamics calculations) have been developed to study the adsorption mechanism on metallic surfaces to forecast corrosion inhibition efficacy which is otherwise quite challenging to determine experimentally. Academicians and scholars are constantly working in the theoretical field to identify the most active adsorption site present within the molecule.

Soltani and co-workers [41] performed quantum chemical calculations to illuminate the adsorption progression of extracts of two gum resins namely, Ferula asafoetida and Dorema ammoniacum in acidic media on mild steel corrosion. Molecular orbitals: highest occupied molecular orbital (HOMO) and lowest unoccupied molecular orbital (LUMO) were generated to recognize the active adsorption sites of gums on steel. The energies of HOMO, LUMO, their energy gap ($\Delta E = E_{HOMO} - E_{LUMO}$), and dipole moment ($\mu$) were calculated to relate their inhibition effect. They found that Ferula asafoetida is more effective corrosion inhibitor than Dorema ammoniacum due to its high E_{HOMO}, lower E_{LUMO} and lower ΔE.

Hasanov *et. al.* [42] considered theoretical calculations to study the effects of poly(N–ethylaniline) as monolayer coating or bilayer coatings on the low carbon steel 1 M H_2SO_4 using AM1 semiempirical method. The theoretical data was well versed with experimental results. It was found that bilayer conducting polymers coatings on the steel surface in 1 M H_2SO_4 are proficient than monolayer coating conditions as interaction of additive with the steel surface was dependent on electron donor tendency of oxidized forms of dimers. Peter *et. al.* [43] used quantitative structure activity relationship method on natural gums to study their efficiency as green corrosion inhibitors.

Egbedi *et al.* [44] implemented theoretical studies on red apple fruit extract to utilize them as green corrosion inhibitors. They observed a dipole interaction between the chemical additive and metal surface and concluded that physical adsorption occurred from electrostatic contact amid charged metal surface and charged centres of additive. Oguzie [45] performed Molecular dynamics simulations to theoretically elucidate the electronic structure as well as adsorption behaviour of alkaloid constituent of Capsicum frutescens extract and to study their corrosion obstructing action on carbon steel.

Umoreon and colleagues [46] investigated DFT–based quantum chemical computations on red apple fruit extract as corrosion inhibition for mild steel in hydrochloric acid (HCl). Inhibition proficiency was found to intensify with the escalation in temperature and concentration. Polarization curves specified red apple extract behaved as mixed–type inhibitor. Low energy band gap ($\Delta E = E_{LUMO} - E_{HOMO}$) indicated a stable complex formation with the metal surface, consequently enhanced adsorption of additive on the metal surface. Also, high dipole moment (μ) value recommend adsorption occurred by physical mechanism.

Mehmeti and Podvorica [47] performed Molecular Dynamic and Monte Carlo calculations to study the adsorption behaviour of carboxylated graphene oxide with reference to Nb, Ta, their oxides (NbO, TaO) surfaces. Shen and coworkers [48], Hajiahmadi and Tavangar [49] also performed theoretical studies on gum Arabic as corrosion inhibitor in oil industry and pyrimidine derivatives corrosion inhibition on iron, respectively.

4.   Sustainable green corrosion inhibitors

Environmental distresses need corrosion inhibitors that are conventional, eco–friendly and non–toxic. Green chemistry provides a reservoir of sustainable green corrosion inhibitors (Fig. 2, Table 3). Plant products are rich sources of globally acceptable environment–friendly green corrosion inhibitors such as alkaloids, phthalocyanines, terpenes, *etc.*

4.1 Natural sources of green corrosion inhibitors

First report on inhibition of corrosion of iron in acids using extracts of gelatin, glue, and bran was given by Marangoni and Stefanelli in 1908. The first patent in the field of corrosion inhibition was given to Baldwin (British Patent 2327), which stated the utilization of naturally obtained molasses with vegetable oils for preserving sheet steel in acids. Recently, gums obtained from plants also finds mentioned as an important class of green corrosion inhibitors. These are mixtures of polysaccharides which either possess hydrophilicity or absorb water and gets swell up on captivating water to form a jelly like structure [43].

Figure 2. Some widely used natural products exploited as green corrosion inhibitors.

Eddy *et al.* [50] researched on gums and recognized them to be harmless, eco–friendly and virtuous green corrosion inhibitors for mild steel in 0.1 M HCl due to various reasons. They can form adsorbate complexes with the metal or their surfaces *via* their functional group and the surface gets blanketed from corrosive agents. The presence of carbohydrates like arbinogalactan, glucoprotein, oligosaccharides, polysaccharides, and sucrose provide ample of oxygen and nitrogen atoms to the gums and hence, possess multiple sites for adsorption. Most of the gums possess carboxylic group, which has property to intensify electronic or charge transfer, hence, enable inhibition *via* adsorption.

Gudić and co-workers [51] implemented Quantum chemical and molecular dynamics studies to evaluate adsorption of one caffeine molecule and polymeric structure of caffeine on copper surface and was well versed with the experimental results. Also, increase in number of caffeine molecules on the adsorptive site increases the binding energy. Caffeine was established as a cathodic inhibitor as a result of adsorption on the copper surface agreeing Langmuir adsorption isotherm. A negative free energy of ~−37 kJ mol^{-1} specifies robust adsorption of the caffeine on copper surface.

The corrosion inhibition efficiency of lemongrass extract was assessed for carbon steel in oilfield water by means of electrochemical impedance spectroscopy as well as weight loss method. Quantum chemical calculations in addition to molecular dynamics simulations were also performed on two abundant species, *namely* geranial and neral present in lemongrass extract to gain knowledge about probable active sites apt for adsorption. The corrosion rate was found to decrease with both geranial and neral, however, former has slightly better tendency to get adsorbed on iron surface than later [52].

The corrosion inhibitory efficacy of Senecio anteuphorbium (SA) extract on S300 steel in 1 M HCl was estimated by using electrochemical impedance spectroscopy and potentiodynamic polarization. SA extract worked as mixed–type inhibitor for steel corrosion in acidic media as established from their polarization curve. The adsorption inhibition was found to enhance with increasing extract concentration and attained a maximum of 91% and followed Langmuir isotherm. Activation energy values proposed it as a physical–chemical adsorption [53].

Sharma *et. al.* [54] investigated anti–corrosion action of Azadirachta indica (neem) for mild steel, aluminium, and tin using computational approach. The reports on NaOH extracts of Ziziphus jujuba leaves [55], stem bark of Mangifera indica [56], Moringa oleifera [56], terminalia arjuna [56], as mixed–type inhibitors for aluminium were also recognized. Hibiscus petal extract in H$_2$SO$_4$ was also testified as a mixed–type inhibitor against pure aluminium [57]. Aspilia africana leaf extract in HCl was cited as a cathodic inhibitor for aluminium [58].

To make environment clean, scientists are constantly exploring natural new products as green corrosion inhibitors. Research work done on this topic includes utilization of barley remains [59], opuntia [60], vernonia [61], hibiscus [62], mangrove [63], tree roots [64], and overabundance of others are explained in Table 3.

Table 3.   *Sustainable green corrosion inhibitors*

Source of natural products	Metal/Alloy	Reference
Leaves and roots of medicinal plants	Mild steel	[64]
Sodium molybdate and calcium gluconate synergism	Carbon steel	[65]
Zinc phosphate base, molybdate base inhibitors	Carbon steel	[66]
Hibiscus sabdariffa	Steel	[62]
Mixture of Garcinia kola + KI	Mild steel in H_2SO_4	[67]
Acid extracts of Calotropis gigantea	Mild steel in acid medium	[68]
Carica papaya extracts	Mild steel in acid medium	[69]
Bacillus mycoides	mild steel, aluminium, zinc	[70]
Structure composed of Zn^{2+}, ascorbate, and $-N, N–bis–$(phosphonomethyl) glycine	Carbon steel	[71]
Phosphonated glycine	Carbon steel	[72]
Rosemary oil	Steel in phosphoric acid	[73]
Powdered seeds of Eugenia jambolans	Mild steel in HCl	[74]
Nicotinic acid	Zn, Zn–Al steels	[75]
Mangrove tannins	Steel	[63]
Carboxymenthylchitosan	Mild steel in 1M HCl	[76]
Datura stramonium extract	Mild steel	[77]
Surfactants	Stainless steel	[78]
Black pepper extract	Mild steel in H_2SO_4	[79]
Cerium	Tin plate	[80]
Azadirachta indica	Mild steel, tin, aluminium	[81]
Xanthates	Steel	[82]
Pomegranate fruit shells	Steel	[83]
Eucalyptus extract	Steel	[84]
Jasminum auriculatum	Steel in $NaCl$	[85]
Barbiturates	Mild steel in 1M HCl	[86]

Table 3.   *Sustainable green corrosion inhibitors contd.....*

Source of natural products	Metal/Alloy	Reference
Tobacco leaves	Mild steel and aluminium	[87]
Natural honey	Carbon steel	[88]
Vanillin	Carbon steel	[89]
Vernonia amygdalina	Aluminium alloy	[61]
Allium cepa	Mild steel	[90]
Guar gum	Carbon steel in acid	[91]
Caffeine	Copper	[51]
Halfabar	Steel	[92]
gum from Dacryodes edulis	Aluminium	[93]
Chitosan	Copper	[94]
Flour and yeast	Iron in acid	[95]
Opuntia extract	Aluminium	[60]
Reducing saccharides fructose and mannose	Aluminium and zinc	[96]
Mangifera indica	Aluminium	[56]
Aspilia	Aluminium	[58]
Ziziphus jujuba leaves	Aluminium in NaOH	[55]
Xylopia aethiopica seeds in KOH	Aluminium	[97]
Moringa oleifera	Aluminium alloy in 1M NaOH	[56]
Eugenol from cloves	Steel	[98]
Polyphosphate–silicate–zinc	Steel	[99]
Recent Patents		
Polybenzoxazine–Chitosan	Steel	[100]
Polyaspartic acid–phosphinocarboxylic acid–hydroxyphosphonoacetic acid	Carbon steel	[101]
Polyaspartic acid–Sn (II) or Sn (IV)	Carbon steel	[102]
Corn stillage product	Steel	[103]

4.2   Ionic liquids (ILs) as green corrosion inhibitors

Scientists are also searching a feasible alternative in ILs as a green option to use them as green corrosion inhibitor apart from inhibitors attained from plants as decontamination of plant extracts is tiresome, strenuous, and tremendously costly as well as more time consuming, and needs comparatively huge quantity of organic solvents which can unfavourably influence the atmosphere and living beings moreover elevated temperature can also degrade the active components and thus reduce comparative inhibition competence. Progress of green chemistry and new greener chemical knowledge presents new synthetic routes for ILs, which is now a day's explored as innovative green corrosion inhibitors. Consequently, there is an urgent need of the hour to expand green inhibitors (due to various applications, Fig. 3) by suitable mean of the amalgamation, which could be accomplished by employing inexpensive and environmentally benign goods. This is the commencement of green chemistry as ILs is sustainable and eco–friendly solvents containing ions that have an ability to relocate a broad variety of organic and inorganic compounds.

ILs pursues the green chemistry principles which are projected by Anastas and his colleague Warner [104]. ILs present enlightening workspace in the era of benign chemistry, thereby gaining immense popularity as they are broadly employed in a range of applications. It is found that chitosan which is a straight chain copolymer type having formula (1–4)–2– acetamido–deoxy–d–glucan (N–acetylglucosamine) and (1–4)–2–amido–2–deoxy–d–glucan (glucosamine), are main basis of ILs. Chitosan is dug out from shell of the crab based on amino polysaccharides and have been utilized to create quite a lot of new–fangled materials [105,106]. Their elevated aptitude for functionalization builds them valid to use in numerous manufacturing applications since they display improved solubility in aqueous and organic solvents as compared to chitosan. Chitosan has a potential to prevent the corrosion of copper in acidic solution since hydrogen chloride is employed as a mixed–type inhibitor (work together with the help of other), pursuing a Langmuir isotherm [107].

Protection of metallic corrosion in the occurrence of ILs engages inhibition of oxidative metallic dissolution at anode along with hydrogen evolution reactions at cathode [108] as shown:

$$M^+X^- \leftrightarrow [(MX)^-]_{ads} \tag{1}$$

$$[(MX)^-]_{ads} + ILsC^+ \leftrightarrow (MX^- \, ILs \, C^+)_{ads} \tag{2}$$

Where

M = dissolution of metals at anode in corrosive solution made in aqueous medium

$ILsC^+$ = ionic liquid cationic equivalent

X^- = ionic liquid anionic equivalent

Mechanistically, anionic counterpart (X^-) draw charge from cationic counterpart of the ILs ($ILsC^+$) *via* Coulombic forces of attraction (known as physio-sorption) and outline a unimolecular layer as hydrophobic composite on the surface of the metal. The accumulation of $ILsC^+$ on the surface of the metal results in the modification of the polarity of the surface, that persuades readsorption of X^- and $ILsC^+$ ions on the surface forming multi-molecular deposits [108]. These deposits then become stable by van der Waal's force of attraction among organic entities of the ILs, consequently forming a more intimately adsorbed layer at the interfaces of metal–electrolyte. In general, $ILsC^+$ network with the surface of the metal and outline the multi-molecular layers, whereas the remaining ILs form hydrophobic surface agglomeration [108,109]. multi-molecular deposits on the ILs separates the metal from its corrosive atmosphere and guards it from degradation.

4.3   Vapor–phase inhibitors (VPI)

VPI are a class of corrosion inhibitors having high atmospheric pressure and are volatile in nature. Now a days they are widely emerging as potent green inhibitors and also requisite in numerous grounds including packaging, electronics, manufacturing processing, resistant solid, finishing, and metal working solutions owing to their harmless characteristics. As they are free from nitrites, phosphates, and halogens hence considered to be completely safe to handle. Moreover, VPI are proficient in preventing the deterioration of iron and non–iron metals.

As literature shows, Bastidas *et al.* [110] considered a range of aspects impacting VPI act and mechanisms. It is also explored that nano–vapor–phase inhibitors derived from renewable agricultural waste stabilize the natural balance of the environment (Trash to Treasure approach). Their properties are dependent on their dispersal, seizing all empty space as well as active sites through defensive vapor molecules which are fascinated to surface of the metal and adsorbed physically or chemically onto the resultant formed nano–protective barrier film [111,112].

5.   How do green inhibitors act as corrosion inhibitors?

Corrosion, an unprompted progression is measured via kinetics studies and is linked to the variation in standard Gibbs free energy ($\Delta G°$). Large $-\Delta G°$ value indicates elevated unprompted progression, *i.e.*, elevated corrosion kinetics [113]. Metals and their alloys tend to form stable corrosion products when exposed to any atmospheric medium [114]. To reduce the corrosion rate, corrosion inhibitors are required. Corrosion products like scale and rust can itself work as corrosion preventers since they are having a tendency to assemble on the metallic surface and serve as outer defensive barriers; though, the corrosion kinetics of the metal is dependent upon the Pilling–Bedworth ratio [115,116]. The ratio could be employed to check the potency of the surface film and is described as follow:

$$\frac{M * d}{n * m * D}$$

where, M, D, n depicts molecular weight, density and amount of the atoms of deposited deterioration product on the surface of metal; d, m represents density and atomic weight of metal, respectively.

If the degree of deterioration of product is reasonably lesser comparative to its precursor metal, then *Md/nmD is less than* 1 and these circumstances institute that the outer surface layer of the deterioration product includes holes and splits that are comparatively non–shielding. However, *Md/nmD greater than 1* specifies that the deterioration of product is higher than that of precursor metal, affirming outer surface layer of the deterioration product as comparatively compact, dense and packed tightly and consequently recognised as a palisade layer.

Generally, these inhibitors include some functionality sometimes along with π–electrons which are in resonance with double or triple bonds, influencing organic compounds through particular binding among required elements *e.g.* nitrogen, oxygen and sulphur *via* freely available lone pairs of electrons, and then get accumulated on the surface of the metal or by providing electrons (e^-) from π–orbitals [117]. Through a mechanistic point of view, process of corrosion can be seen *via* following way: During deterioration, metal ions (M^+) flow towards active sites of fluid (anode) and transfer e^- towards less active site *i.e.* cathodic acceptor. This cathodic route needs the availability of an e^- acceptor like hydrogen ions (H^+), oxygen or other oxidizing agents, or. Deterioration could be reduced by delaying or entirely ending the cathodic and/or anodic reactions. A protective barrier is formed when an inhibitor is adsorbed on the metallic surface. This barrier interacts with cathodic and/or anodic reaction areas to reduce the oxidation reactions or other corrosive reactions [22].

Common cathodic reactions occurring in corrosive environment are as follow:

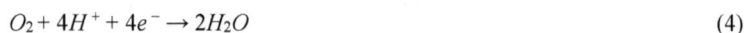

$$2H^+ + 2e^- \rightarrow H_2 \tag{3}$$

$$O_2 + 4H^+ + 4e^- \rightarrow 2H_2O \tag{4}$$

Reduction reaction comprises evolution of hydrogen gas as shown:

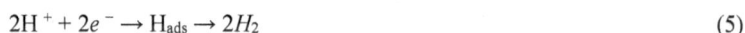

$$2H^+ + 2e^- \rightarrow H_{ads} \rightarrow 2H_2 \tag{5}$$

Eq. 5 shows accumulation of hydrogen ions (H^+) on the surface of metal and lead to evolution of hydrogen gas (H_2) on cathodic surface due to the amalgamation with other hydrogen ions (H^+). Quantity of hydrogen gas evolved refers to capacity of inhibitor to stop this process and saving metal from deterioration [23]. Here adsorption of inhibitor on the exposed surface of metal occurred as neutral molecules.

This process can easily be explained by following reaction:

$$\text{Inhibitor} + n\,H_{ads} \rightarrow \text{Inhibitor}_{ads} + H_2 \qquad\qquad (6)$$

or *via* replacement of molecules of water on the surface of metal, as explained using Eq. 4.

These green corrosion inhibitors possess wide variety of applications in various fields and are displayed in Fig. 3.

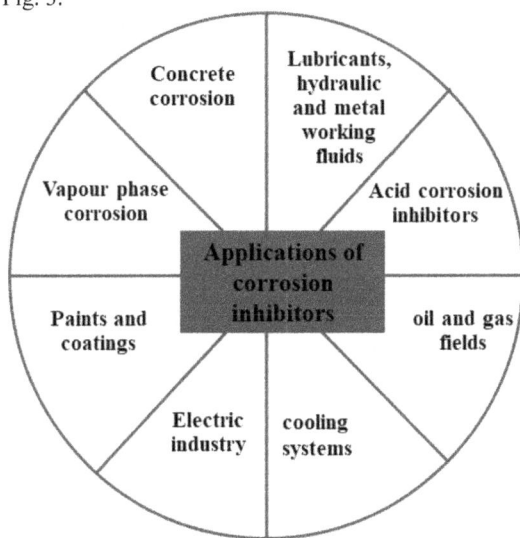

Figure 3.   *Applications of green corrosion inhibitors in various industries.*

Conclusion

The goal of this chapter is to reassess biogenic compounds as valuable green and natural corrosion inhibitors due to various characteristics such as their biodegradability, effortless accessibility, economical and safe nature. Literature surveys mentioned biogenic plant extracts as efficient green deteriorators for different metals and their alloys. Moreover, the article also centred on "the competence of green and sustainable inhibitors for the corrosion of metals and alloys is favoured as compare to conventional corrosion inhibitors". At last, interest was shown to the adsorption behaviour of green inhibitors usually, following the Langmuir isotherm and by means of DFT–based quantum chemical simulations. Quantum chemical computations offer a superior approach interested in the inhibition pathway. The affinity to investigate VPI and green nano inhibitors is a fresh quarter for upcoming investigations.

Voluminous research in this field is yet to be explored, particularly virtual computational modelling of the chief extract entities of diverse metals and their alloys. additional investigation should be determined too on various plant extraction protocols and their dynamic components along with scale–up procedures for business relevance that are required to market these biogenic extracts to successfully reinstate traditional chemicals.

References

[1] B. Cwalina, Understanding biocorrosion fundamentals and applications, in: T. Liengen, D. Féron, R. Basséguy, I.B. Beech (Eds.), Biodeterioration of Concrete, Brick and other Mineral–Based Building Materials, Woodhead Publishing, 2014, pp. 281–312. https://doi.org/10.1533/9781782421252.3.281

[2] V.S. Sastri, Green Corrosion Inhibitors: Theory and Practice, John Wiley & Sons, Inc., Hoboken, New Jersey, 2011. https://doi.org/10.1002/9781118015438

[3] Z. Petrović, Catastrophes Caused by Corrosion, Military Technical Courier, 64(4) (2016) 1048–1064. https://doi.org/10.5937/vojtehg64-10388

[4] A. Javed, S. Pal, E.K. Krishnan, P. Sahni, T.K. Chattopadhyay, Surgical management and outcomes of severe gastrointestinal injuries due to corrosive ingestion, World J. Gastrointest. Surg. 4(5) (2012) 121–125. https://doi.org/10.4240/wjgs.v4.i5.121

[5] A. Bennett, J.R. Goldblum, R.D. Odze, Inflammatory disorders of the esophagus, in: R. Odze, J. Goldblum (Eds.), Surgical Pathology of the GI Tract, Liver, Biliary Tract, and Pancreas, Saunders, 2009, pp. 231–267. https://doi.org/10.1016/B978-141604059-0.50014-X

[6] S.M. Keh, N. Onyekwelu, K. McManus, J. McGuigan, Corrosive injury to upper gastrointestinal tract: Still a major surgical dilemma, World J. Gastrointest. Surg. 12(32) (2006) 5223–5228.

[7] R.E. Allen, M.J. Thoshinsky, R.J. Stallone, T.K. Hunt, Corrosive Injuries of the Stomach, Arch. Surg. 100(4) (1970) 409–413. https://doi.org/10.1001/archsurg.1970.01340220085015

[8] T.P. Hoar, Report of the committee on corrosion and protection: a survey of corrosion and protection in the United Kingdom, London HMSO, UK, 1971.

[9] D. Behrens, Research and development programme on 'Corrosion and Corrosion Protection' in the German federal republic, Br. Corros. J. 10(3) (1975) 122–127. https://doi.org/10.1179/000705975798320657

[10] R. Bhaskaran, L. Bhalla, A. Rahman, An Analysis of the Updated Cost of Corrosion in India, Mater. Perform. 53(8) (2014) 56-65.

[11] Committee on Corrosion and Protection, Corros. Eng. Jpn. 26(7) (1977) 401.

[12] Y.M. Panchenko, A.I. Marshakov, L.A. Nikolaeva, V.V. Kovtanyuk, T.N. Igonin, T.A. Andryushchenko, Comparative estimation of long-term predictions of corrosion losses for carbon steel and zinc using various models for the Russian territory, Corros. Eng. Sci. Technol. 52(2) (2017) 149–157. https://doi.org/10.1080/1478422X.2016.1227024

[13] H. Wall, L.Wadsö, Sheet pile corrosion in Swedish Harbours - An inventory of corrosion surveys along the swedish coast, in: European Corrosion Congress 2011 (EUROCORR 2011), pp. 2520.

[14] K.S. Rajagopalan, Report on metallic corrosion and its prevention in India, The Hindu, Madras, India, Nov 12, 1973.

[15] G.H. Koch, M.P.H. Brongers, N.G. Thompson, Cost of corrosion and prevention strategies in the united states, C. C. Technologies Laboratories, Inc., Dublin, Ohio, USA, 2001.

[16] Assessment of Global Cost of Corrosion, 2015. http://impact.nace.org/documents/appendix–a.pdf (accessed 30 December 2019).

[17] O.S. Shehata, L.A. Korshed, A. Attia, Green corrosion inhibitors, past, present, and future, in: M. Aliofkhazraei (Ed.), Corrosion Inhibitors, Principles and Recent Applications, InTech Publishing Inc., Croatia, 2018, pp. 121–142. https://doi.org/10.5772/intechopen.72753

[18] G.T. Hefter, N.A. North, S.H. Tan, Organic corrosion inhibitors in neutral solutions; Part 1–Inhibition of steel, copper, and aluminum by straight chain carboxylates, Corros. 53(8) (1997) 657–667. https://doi.org/10.5006/1.3290298

[19] M. Bethencourt, F.J. Botana, J.J. Calvino, M. Marcos, M.A. Rodriguez–Chacon Lanthanide compounds as environmentally–friendly corrosion inhibitors of aluminium

alloys: A review, Corros. Sci. 40(11) (1998) 1803–1819.
https://doi.org/10.1016/S0010-938X(98)00077-8

[20] A.A.F Sabirneeza, R. Geethanjali, S. Subhashini, Polymeric corrosion inhibitors for iron and its alloys: a review, Chem. Eng. Comm. 202(22) (2015) 232–244. https://doi.org/10.1080/00986445.2014.934448

[21] K. Krishnaveni, J. Ravichandran, Effect of aqueous extract of leaves of morinda tinctoria on corrosion inhibition of aluminium surface in HCl medium, T. Nonferr. Metal. Soc. China 24(8) (2014) 2704–2712. https://doi.org/10.1016/S1003-6326(14)63401-4

[22] P. Singh, V. Srivastava, M.A. Quraishi, Novel quinoline derivatives as green corrosion inhibitors for mild steel in acidic medium: electrochemical, SEM, AFM, and XPS studies, J. Mol. Liq. 216(1) (2016) 164–173. https://doi.org/10.1016/j.molliq.2015.12.086

[23] E.E. Oguzie. Inhibition of acid corrosion of mild steel by telfaria occidentalis extract, Pigm. Resin Technol. 34(6) (2005) 321–326. https://doi.org/10.1108/03699420510630336

[24] P. Zarras, J.D. Stenger–Smith, Smart inorganic and organic pretreatment coatings for the inhibition of corrosion on metals/alloys, in: A. Tiwari, J. Rawlins, L.H. Hihara (Eds.), Intelligent Coatings for Corrosion Control, Oxford, UK, 2014, pp. 59–91. https://doi.org/10.1016/B978-0-12-411467-8.00003-9

[25] M. Chigondo, F. Chigondo, Recent Natural corrosion inhibitors for mild steel: an overview, J. Chem. 2016 (2016) 1–7. https://doi.org/10.1155/2016/6208937

[26] A. Bonfanti, C. Lecomte, B.J. Little, J.S. Lee, Bioactive environments: corrosion, in: S. Hashmi (Ed.), Reference Module in Materials Science and Materials Engineering, 2018, pp. 533–537. https://doi.org/10.1016/B978-0-12-803581-8.02888-5

[27] K. Sieradzki, Dealloying, in: K.H.J. Buschow, R. Cahn, M. Flemings, B. Ilschner, E.Kramer, S. Mahajan, P. Veyssier (Eds.), Encyclopaedia of Materials: Science and Technology, Pergamon, 2001, pp. 1984–1985.

[28] E.L. Colvin, Aluminum Alloys: Corrosion, in: K.H.J. Buschow, R. Cahn, M. Flemings, B. Ilschner, E.Kramer, S. Mahajan, P. Veyssier (Eds.), Encyclopaedia of Materials: Science And Technology, Pergamon, 2001, 107–110). https://doi.org/10.1016/B0-08-043152-6/00022-X

[29] D.Y. Li, Corrosive Wear, in: Q.J. Wang, Y.W. Chung (Eds.), Encyclopaedia of Tribology, Springer, Boston, MA, 2013, pp. 19.

[30] N.D. Nam, P.V. Hien, N. T. Hoai, V.T.H. Thu, A study on the mixed corrosion inhibitor with a dominant cathodic inhibitor for mild steel in aqueous chloride solution, J. Taiwan Inst. Chem. E. 91 (2018), 556–569. https://doi.org/10.1016/j.jtice.2018.06.007

[31] G.M. Spinks, A.J. Dominis, G.G. Wallace, D.E. Tallman, Electroactive conducting polymers for corrosion control, J. Solid State Electrochem. 6 (2002) 85–100. https://doi.org/10.1007/s100080100211

[32] J.R. Santos, L.H.C. Mattoso, A.J. Motheo, Investigation of corrosion protection of steel by polyaniline films, Electrochim. Acta. 43 (1998) 309–313. https://doi.org/10.1016/S0013-4686(97)00052-2

[33] C.A. Mann, B.E. Lauer, C.T. Hultin, Organic inhibitors of corrosion, aliphatic amines, Ind. Eng. Chem. 28(2) (1936) 159–163. https://doi.org/10.1021/ie50314a004

[34] N. Hackerman, Effect of sorption on metal dissolution in aqueous acid solution, Trans NY Acad Sci. 17(1) (1954) 7–11. https://doi.org/10.1111/j.2164-0947.1954.tb01331.x

[35] H. Kaesche, N. Hackerman, Corrosion Inhibition by Organic Amines, J. Electrochem. Soc. 105(4) (1958) 191–198. https://doi.org/10.1149/1.2428796

[36] A. Popova, E. Sokolova, S. Raicheva, M. Christov, AC and DC study of the temperature effect on mild steel corrosion in acid media in the presence of benzimidazole derivatives, Corros. Sci. 45 (2003) 33–58. https://doi.org/10.1016/S0010-938X(02)00072-0

[37] A. Popova, M. Christov, S. Raicheva, E. Sokolova, Adsorption and inhibitive properties of benzimidazole derivatives in acid mild steel corrosion, Corros. Sci. 46 (2004) 1333–1350. https://doi.org/10.1016/j.corsci.2003.09.025

[38] A. Popova, M. Christov, A. Zwetanova, Effect of the molecular structure on the inhibitor properties of azoles on mild steel corrosion in 1 M hydrochloric acid, Corros. Sci. 49 (2007) 2131–2143. https://doi.org/10.1016/j.corsci.2006.10.021

[39] K. Babic–Samardzija, C. Lupu, N. Hackerman, A.R. Barron, Inhibitive properties, adsorption and surface study of butyn–1–ol and pentyn–1–ol alcohols as corrosion inhibitors for iron in HCl, J. Mater. Chem. 15 (2005) 1908–1916. https://doi.org/10.1039/b416202a

[40] M. Finšgar, J. Jackson, Application of corrosion inhibitors for steels in acidic media for the oil and gas industry: a review, Corros. Sci. 86 (2014) 17–41. https://doi.org/10.1016/j.corsci.2014.04.044

[41] M. Behpour, S.M. Ghoreishi, M. Khayatkashani, N. Soltani, The effect of two oleo–gum resin exudate from Ferula asafoetida and Dorema ammoniacum on mild steel corrosion in acidic media, Corros. Sci. 53 (2011) 2489–2501. https://doi.org/10.1016/j.corsci.2011.04.005

[42] R. Hasanov, S. Bilgiç, G. Gece, Experimental and theoretical studies on the corrosion properties of some conducting polymer coatings, J. Solid State Electrochem. 15 (2011) 1063–1070. https://doi.org/10.1007/s10008-010-1275-6

[43] A. Peter, I. B. Obot, Sanjay K. Sharma, Use of natural gums as green corrosion inhibitors: an overview, Int. J. Ind. Chem. 6 (2015) 153–164. https://doi.org/10.1007/s40090-015-0040-1

[44] N.O. Obi–Egbedi, I.B. Obot, S.A. Umoren, Experimental and theoretical studies of red apple fruit extract as green corrosion inhibitor for mild steel in HCl solution, Arab. J. Chem. 5 (2012) 361–373. https://doi.org/10.1016/j.arabjc.2010.09.002

[45] E.E. Oguzie, K.L. Oguzie, C.O. Akalezi, I.O. Udeze, J.N. Ogbulie, V.O. Njoku Natural products for materials protection: corrosion and microbial growth inhibition using capsicum frutescens biomass extracts, ACS Sustain. Chem. Eng. 1 (2013) 214–225. https://doi.org/10.1021/sc300145k

[46] S. Umoren, I.B. Obot, Z. Gasem, N.A. Odewunmi, Experimental and theoretical studies of red apple fruit extract as green corrosion inhibitor for mild steel in HCl solution, J. Disper. Sci. Technol. 36(6) (2015) 789–802. https://doi.org/10.1080/01932691.2014.922887

[47] M. Valbonë, F.I. Podvorica, Experimental and theoretical studies on corrosion inhibition of niobium and tantalum surfaces by carboxylated graphene oxide, Materials 11 (2018) 893–905. https://doi.org/10.3390/ma11060893

[48] C. Shen, V. Alvarez, J.D.B. Koenig, Jing–Li Luo, Gum Arabic as corrosion inhibitor in the oil industry: experimental and theoretical studies, Corros. Eng. Sci. Tech., 54 (2019) 444–454. https://doi.org/10.1080/1478422X.2019.1613780

[49] Z. Hajiahmadi, Z. Tavangar, Extensive theoretical study of corrosion inhibition efficiency of some pyrimidine derivatives on iron and the proposal of new inhibitor, J. Mol. Liq. 284 (2019) 225–231. https://doi.org/10.1016/j.molliq.2019.04.009

[50] N.O. Eddy, C.E. Gimba, P.O. Ameh, E.E. Ebenso, GCMS studies on Anogessus leocarpus (Al) gum and their corrosion inhibition potential for mild steel in 0.1 M HCl, Int. J. Electrochem. Sci. 6 (2011) 5815–5829.

[51] S. Gudić, E.E. Oguzie, A. Radonić, L. Vrsalović, I. Smoljko, M. Kliškić, Inhibition of copper corrosion in chloride solution by caffeine isolated from black tea, Maced. J. Chem. Chem. En. 33(1) (2014) 13–25. https://doi.org/10.20450/mjcce.2014.441

[52] M.A. Deyab, M.M. Osman, A.E. Elkholya, F. El–Taib Heakal, Green approach towards corrosion inhibition of carbon steel in produced oilfield water using lemongrass extract, RSC Adv. 7 (2017) 45241–45251. https://doi.org/10.1039/C7RA07979F

[53] R. Idouhli, Y. Koumya, · M. Khadiri, A. Aityoub, A. Abouelfida, A. Benyaich, Inhibitory effect of Senecio anteuphorbium as green corrosion inhibitor for S300 steel, Int. J. Indus. Chem. 10 (2019) 133–143. https://doi.org/10.1007/s40090-019-0179-2

[54] S.K. Sharma, A. Peter, I.B. Obot, Potential of Azadirachta indica as a green corrosion inhibitor against mild steel, aluminum, and tin: a review, J. Anal. Sci. Tech. 6 (2015) 26–41. https://doi.org/10.1186/s40543-015-0067-0

[55] A. Sirajunnisa, M.I. Fazal Mohamed, A. Subramania, B.R. Venkatraman, Eur. J. Appl. Sci. Tech. 1 (2014) 23–31.

[56] N. Chaubey, Savita, V.K. Singh, M.A. Quraishi, Corrosion inhibition performance of different bark extracts on aluminium in alkaline solution, J. Assoc. Arab. Univ. Basic Appl. Sci. 22 (2016) 38–44. https://doi.org/10.1016/j.jaubas.2015.12.003

[57] L.B. Molina–Ocampo, M.G. Valladares–Cisneros, J.G. Gonzalez–Rodriguez, Int. J. Electrochem. Sci. 10 (2015) 388–403.

[58] I.M. Mejeha, M. C. Nwandu, K. B. Okeoma, L. A. Nnanna, M. A. Chidiebere, F. C. Eze and E. E. Oguzie, J. Mater. Sci. 47 (2012) 2559–2572. https://doi.org/10.1007/s10853-011-6079-2

[59] M. Saadawy, An important world crop – Barley – As a new green inhibitor for acid corrosion of steel, Anti-Corros. Methods M. 62(4) (2015) 220–228. https://doi.org/10.1108/ACMM-12-2013-1333

[60] A.Y. El–Etre, Inhibition of aluminum corrosion using Opuntia extract, Corros. Sci. 45(11) (2003) 2485–2495. https://doi.org/10.1016/S0010-938X(03)00066-0

[61] G.O. Avwiri, F.O. Igho, Inhibitive action of Vernonia amygdalina on the corrosion of aluminium alloys in acidic media, Mat. Lett. 57(22) (2003) 3705–3711. https://doi.org/10.1016/S0167-577X(03)00167-8

[62] E.E. Oguzie, Corrosion inhibitive effect and adsorption Behaviour of Hibiscus Sabdariffa extract on mild steel in acidic media, Port. Electrochim. Acta 26 (2008) 303–314. https://doi.org/10.4152/pea.200803303

[63] A. Rahim, E. Rocca, J. Steinmetz, M.J. Kassim, R. Adnan, M.S. Ibrahim, Mangrove tannins and their flavanoid monomers as alternative steel corrosion inhibitors in acidic medium, Corros. Sci., 49(2) (2004) 402–417. https://doi.org/10.1016/j.corsci.2006.04.013

[64] D. Bouknana, B. Hammouti, H.S. Caid, S. Jodeh, A. Bouyanzer, A. Aouniti, I. Warad, Aqueous extracts of olive roots, stems, and leaves as eco–friendly corrosion inhibitor for steel in 1M HCl medium, Int. J. Ind. Chem. 6 (2015) 233–245. https://doi.org/10.1007/s40090-015-0042-z

[65] S.M.A Shibli, V.A. Kumary, Inhibitive effect of calcium gluconate and sodium molybdate on carbon steel, Anti–Corros. Method Mater., 51(4) (2004) 277–281. https://doi.org/10.1108/00035590410541355

[66] A.M. Alfuraij, Corrosion and Prevention—2000, Auckland, NewZealand, 2000, pp. 19–22.

[67] P.C. Okafor, V.I. Osabor, E.E. Ebenso, Eco–friendly corrosion inhibitors: inhibitive action of ethanol extracts of Garcinia kola for corrosion of mild steel in H_2SO_4 solutions, Pigm. Resin Technol. 36 (2007) 299–305. https://doi.org/10.1108/03699420710820414

[68] S. Kumar, Eco–friendly corrosion inhibitors: Synergistic effect of ethanol extracts of calotropis for corrosion of mild steel in acid media using mass loss and thermometric technique at different temperatures, Prot. Met. Phys. Chem. Surf. 52 (2016) 376–380. https://doi.org/10.1134/S2070205116020167

[69] P.C. Okafor, E.E. Ebenso, Inhibitive action of carica papaya extracts on the corrosion of mild steel in acidic media and their adsorption characteristics, Pigm. Resin Technol. 36 (2007) 134–140. https://doi.org/10.1108/03699420710748992

[70] E Juzeliunas, R Ramanauskas, Influence of wild strain Bacillus mycoides on metals: From corrosion acceleration to environmentally friendly protection Electrochim. Acta 51(27) (2006) 6085–6090. https://doi.org/10.1016/j.electacta.2006.01.067

[71] B.V.A. Rao, S. S. Rao, M.V. Rao, Environmentally friendly ternary inhibitor formulation based on N, N–bis (phosphonomethyl) glycine, Corros. Eng. Sci. Technol. 43(1) (2008) 46–53. https://doi.org/10.1179/174327807X214635

[72] B.V.A. Rao, M.V. Rao, Long Term Prediction and Modelling of Corrosion, Eurocorr 2004, Nice, France, Sept 12–16, 2004.

[73] M. Bendahou, M. Benabdellah, B. Hammouti, A study of rosemary oil as a green corrosion inhibitor for steel in 2 M H_3PO_4, Pigm. Resin Technol. 35(2) (2006) 95–100. https://doi.org/10.1108/03699420610652386

[74] S. A. Verma, G. N. Mehta, Effect of acid extracts of powdered seeds of eugenia jambolans on corrosion of mild steel in HCl–study by DC polarisation techniques, Trans SAEST, 32(4) (1997) 89–93.

[75] H. Ju, Y. Li, Nicotinic acid as a nontoxic corrosion inhibitor for hot dipped Zn and Zn–Al alloy coatings on steels in diluted hydrochloric acid, Corros. Sci. 49(11) (2007) 4185–4201. https://doi.org/10.1016/j.corsci.2007.05.015

[76] S. Cheng, S. Chen, T. Liu, X. Chang, Y. Yin, Carboxymenthylchitosan as an ecofriendly inhibitor for mild steel in 1 M HCl, Mater. Lett. 61 (2007) 3276–3280. https://doi.org/10.1016/j.matlet.2006.11.102

[77] P.B. Raja, M.G. Sethuraman, Studies on the inhibitive effect of Datura stramonium extract on the acid corrosion of mild steel, Surf. Rev. Lett. 14(6) (2007) 1157–1164. https://doi.org/10.1142/S0218625X07010743

[78] Z. Wei, P. Duby, P. Somasundaran, Corrosion 2004, New Orleans, LA, 2004, pp. 14.

[79] P.B. Raja, M.G. Sethuraman, Inhibitive effect of black pepper extract on the sulphuric acid corrosion of mild steel, Mater. Lett. 62 (2008) 2977–2979. https://doi.org/10.1016/j.matlet.2008.01.087

[80] M.A. Arenas, A. Conde, J.J. Damborenea, Cerium: a suitable green corrosion inhibitor for tinplate, Corros. Sci. 44(3) (2002) 511–520. https://doi.org/10.1016/S0010-938X(01)00053-1

[81] S.K. Sharma, A. Peter, I.B. Obot, Potential of Azadirachta indica as a green corrosion inhibitor against mild steel, aluminum, and tin: a review, J. Anal. Sci. Tech. 6 (2015) 26–41. https://doi.org/10.1186/s40543-015-0067-0

[82] V.A. Altekar, I. Singh, M.K. Banerjee, M.N. Singh, T.R. Soni, Proceedings of the 5th European Symposium on Corrosion Inhibitors, Vol. University of Ferrara, Italy, 2 (1980) 367.

[83] A.A. El–Hosary, R.M. Saleh, H.A. El–Dahan, Proceedings of the 7th European Symposium on Corrosion Inhibitors, University of Ferrara, Italy, 1 (1990) 725.

[84] A. Minhaj, P.A. Saini, M.A. Quraishi, I.H. Farooqi, A Study of natural compounds as corrosion inhibitors for industrial cooling systems, Corros. Prev. Control 42(6) (1999) 32–38.

[85] I.H. Farooqi, M.A. Quraishi, P.A. Saini, Corrosion prevention of mild steel in 3% NaCl water by some naturally–occurring substances, Corros. Prev. Control 70 (1999) 370–372.

[86] G. Kardas, R. Solmaz, Electrochemical investigation of barbiturates as green corrosion inhibitors for mild steel protection, Corros. Rev. 24 (2006) 151–171. https://doi.org/10.1515/CORRREV.2006.24.3-4.151

[87] K. Srivastav, P. Srivastava, Studies on plant materials as corrosion inhibitors. Br. Corros. J. 16(4) (1981) 221–223. https://doi.org/10.1179/000705981798274788

[88] A. Y. El–Etre, M. Abdallah, Natural honey as corrosion inhibitor for metals and alloys. II. C–steel in high saline water, Corros. Sci. 42(4) (2000)731–738. https://doi.org/10.1016/S0010-938X(99)00106-7

[89] A.Y. El–Etre, Inhibition of acid corrosion of aluminum using vaniline, Corros. Sci. 43(6) (2001)1031–1039. https://doi.org/10.1016/S0010-938X(00)00127-X

[90] K.S. Parikh, K. J. Joshi, Natural compounds onion (Allium cepa), garlic (Allium sativum) and bitter gourd (Momordica charantia) as corrosion inhibitors for mild steel in hydrochloric acid, Transactions—Soc. Adv. Electrochem. Sci. Tech. 39(1/2) (2004) 29–35.

[91] M. Abdallah, Guar gum as corrosion inhibitor for carbon steel in sulfuric acid solutions, Port. Electrochim. Acta. 22(2) (2004) 161–175. https://doi.org/10.4152/pea.200402161

[92] A.M. Abdel–Gaber, B.A. Abd–El–Nabey, I.M. Sidahmed, A.M. El–Zayady, M. Saadawy, Inhibitive action of some plant extracts on the corrosion of steel in acidic media, Corros. Sci. 48(9) (2006) 2765–2779. https://doi.org/10.1016/j.corsci.2005.09.017

[93] S.A. Umoren, I.B. Obot, E.E. Ebenso, N. Obi–Egbedi, Studies on the inhibitive effect of exudate gum from Dacroydes edulis on the acid corrosion of aluminium, Port. Electrochim. Acta 26(2) (2008)199–209. https://doi.org/10.4152/pea.200802199

[94] M.N. El–Haddad, Chitosan as a green inhibitor for copper corrosion in acidic medium, Int. J. Biol. Macromol. 55(2013) 142–149.

[95] J. Baldwin, British Patent. 1895:2327.
https://doi.org/10.1016/j.ijbiomac.2012.12.044

[96] B. Müller, Corrosion inhibition of aluminium and zinc pigments by saccharides.
Corros. Sci. 44(7) (2002)1583–1591. https://doi.org/10.1016/S0010-938X(01)00170-6

[97] O.F. Nwosu, E. Osarolube, Corrosion inhibition of aluminium alloy in 0.75 M KOH
alkaline solution using xylopia aethiopica seed extract, Phys. Rev. Res. Int. 4 (2014)
1235–1243. https://doi.org/10.9734/PSIJ/2014/9928

[98] E. Chaieb, A. Bouyanzer, B. Hammouti, M. Benkaddour, Inhibition of the corrosion
of steel in 1M HCl by eugenol derivatives, Appl. Surf. Sci. 246(1) (2005) 199–206.

[99] M. Islam, H. Al–Mazeedi, A.M. Abdullah, 8th International Congress on Metallic
Corrosion, Vol. II, Mainz, West Germany, 1981, pp. 1233–1238.
https://doi.org/10.1016/j.apsusc.2004.11.011

[100] T.B. Hafiz, A.A. Almathami, G. Agrawal, M. Hoegerl, Green high–efficiency
corrosion inhibitor, 2018, US Patent 057947.

[101] M. Drewniak, L.H. Steimel, Composition and method for inhibiting corrosion and
scale, 2017. US Patent. 0306506A1.

[102] D.L. Erickson, R.A. Johnson, M.R. LaBrosse, P.R. Young, Corrosion inhibiting
methods, 2016, US Patent 9290850B2.

[103] M. Kharshan, A. Fursan, K. Gillette, R. Kean, Bio–based corrosion inhibitor, 2013,
US Patent 8409340B1.

[104] P.T. Anastas, J.C. Warner, Green Chemistry: Theory and Practice, New York,
Oxford University Press, 2000.

[105] N.M. Sarbon, S. Sandanamsamy, S.F. Kamaruzaman, F. Ahmad, Chitosan
extracted from mud crab (Scylla olivicea) shells: physicochemical and antioxidant
properties, J. Food Sci. Technol. 52(7) (2015) 4266–4275.
https://doi.org/10.1007/s13197-014-1522-4

[106] Z. Yao, C. Zhang, Q. Ping, L. Yu, A series of novel chitosan derivatives: synthesis,
characterization and micellar solubilization of paclitaxel, Carbohydr. Polym. 68(4)
(2007) 781–792. https://doi.org/10.1016/j.carbpol.2006.08.023

[107] M.N. El–Haddad, Chitosan as a green inhibitor for copper corrosion in acidic
medium, Int. J. Biol. Macromol. 55 (2013) 142–149.
https://doi.org/10.1016/j.ijbiomac.2012.12.044

[108] N.V. Likhanova, M. A. Domínguez–Aguilar, O. Olivares–Xometl, N. Nava–Entzana, E. Arce, H. Dorantes, The effect of ionic liquids with imidazolium and pyridinium cations on the corrosion inhibition of mild steel in acidic environment, Corros. Sci. 52(6) (2010) 2088–2097. https://doi.org/10.1016/j.corsci.2010.02.030

[109] R. Fuchs–Godec, The adsorption, CMC determination and corrosion inhibition of some N–alkyl quaternary ammonium salts on carbon steel surface in 2M H_2SO_4, Colloid. Surf. A. 280(1) (2006) 130–139. https://doi.org/10.1016/j.colsurfa.2006.01.046

[110] D.M. Bastidas, E. Cano, E.M. Mora, Volatile corrosion inhibitors: a review. Anti–Corros. Method M. 52(2) (2005) 71–77. https://doi.org/10.1108/00035590510584771

[111] A. Subramanian, M. Natesan, V.S. Muralidharan, K. Balakrishnan, T. Vasudevan An overview: vapor phase corrosion inhibitors, Corros. 56(2) (2000) 144–155. https://doi.org/10.5006/1.3280530

[112] E. Vuorinen, E. Kálmán, W. Focke Introduction to vapour phase corrosion inhibitors in metal packaging, Surf. Eng. 20(4) (2004) 281–284. https://doi.org/10.1179/026708404225016481

[113] S. Manimegalai, P. Manjula, Thermodynamic and adsorption studies for corrosion inhibition of mild steel in aqueous media by Sargasam swartzii (brown algae), J. Mater. Environ. Sci. 6(6) (2015) 1629–1637.

[114] Y. Roh, S.Y. Lee, M.P. Elless, Characterization of corrosion products in the permeable reactive barriers, Environ. Geol. 40(1–2) (2000) 184–194. https://doi.org/10.1007/s002540000178

[115] C.H. Xu, W. Gao, Pilling–Bedworth ratio for oxidation of alloys, Mater. Res. Innov. 3(4) (2000) 231–235. https://doi.org/10.1007/s100190050008

[116] R.E. Bedworth, N.B. Pilling, The oxidation of metals at high temperatures, Inst. Metals, 29(3) (1923) 529–582.

[117] O.S. Shehata, L.A. Khorshed, H.S. Mandour, Effect of acetamide derivative and its Mn–complex as corrosion inhibitor for mild steel in sulphuric acid, Egypt. J. Chem. 60(2) (2017) 243–259.

Materials Research Forum LLC
https://doi.org/10.21741/9781644901496-8

Chapter 8

Sustainable Corrosion Inhibitors for Copper and its Alloys

B. El Ibrahimi*

Department of Chemistry, Faculty of Sciences. IBN ZOHR University. Agadir. Morocco

* brahimmhm@gmail.com

Abstract

The good properties of copper, as well as its alloys, make them a often used metallic materials in various industries. Regardless of their excellent corrosive resistance, the corrosion process of copper materials can occur under some specific conditions, hence, the need for corrosion inhibitors. Recently, due to many environmental agencies, the "green" aspect was introduced in many fields, including inhibition of corrosion. Keeping in mind economic and eco-friendly aspects, a wide range of compounds were employed as ecological inhibitors for copper materials. For this purpose, the current chapter aims to explore the application of numerous compounds as sustainable inhibitors to control the corrosion of copper materials in various media.

Keywords

Copper, Alloy, Corrosion, Sustainable Inhibitor, Biopolymer, Plant Extract, Ionic Liquids, Amino Acid, Acid, Alkaline, Ecological, Green

Contents

1.   Introduction

Metal corrosion or metal degradation is a spontaneous and natural phenomenon that converts it into more stable forms (natural forms) like oxides, sulfide and so on. This natural conversion occurs via the electrochemical and/or chemical transformations caused by the adjacent environments. Metallic corrosion affects different fields such as railways bridges, industries, buildings, traffic and households [1]. In a recent study [2] performed by NACE[1], about 2.5 trillion $ (USA) is the value of economic cost caused by the corrosion phenomena. Fig. 1 illustrates in detail the economic costs of corrosion. Nevertheless, this loss can be reduced up to 15–35%, if the corrosion protection approaches are applied.

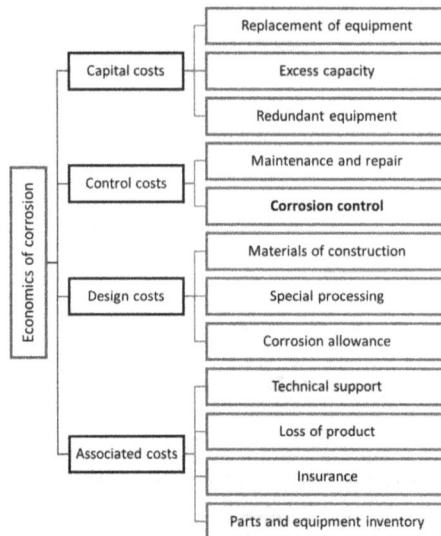

Figure 1.   The economic costs of corrosion [10].

---

[1] National Association of Corrosion Engineers.

To overcome such disadvantageous phenomena or reduce its impact, a variety of protection technologies were used, among them using a corrosion inhibitor is the most attractive method and practical one [3], because it is a rentable economic solution owing to its simplicity of application and its high efficiency. By definition, an inhibitor of corrosion is a chemical compound that introduced to aggressive medium to result in the decline of metal dissolution to an acceptable level [4,5].

Based on their chemical composition, the inhibitors of corrosion could be categorized into two major classes, i.e. organic inhibitors and inorganic inhibitors. The first set of inhibitors is wieldy applied in acidic and sometimes in near-neutral media, whereas the second inhibitors class is used for metal protection in alkaline environments. Since the first apparition of organic compounds to be used as corrosion inhibitors in 1945[2] [6], the developments in this field undergoing outstanding development. Several inhibitors with high anti-corrosion efficiencies were synthesized and reported, like chromates-based species. In this context, the effect of those compounds on human health and/or surrounding environment is missed and is not discussed. Recently, due to the activities of many environmental agencies worldwide, the "green" aspect was introduced in almost fields, in which the inhibition of corrosion is concerning [7]. Consequently, the improvement of new inhibitors of corrosion without or with negligible undesirable environmental impacts (i.e. sustainable corrosion inhibitors) is considered to be more desired [8,9].

Keeping in mind economic criteria and eco-friendly aspects together, a wide range of compounds were reported and used as potent sustainable inhibitors for several metallic materials in different corrosive environments. According to available literature, sustainable corrosion inhibitors can be subdivided into two different classes, the first one is the organic inhibitors, which regroups: amino acids, ionic liquids, biopolymers, plant extracts, surfactants, and pharmaceutical compounds. Concerning the second class, it contains the rare earth compounds, which are the inorganic inhibitors [10].

The attractive chemical and physical properties of copper, as well as its alloys, make them the largely used metals in various industrial fields, such as chemical industries, heating, electronics industries and cooling structures, other than in daily life [11-14]. Regardless of their excellent corrosive resistance, the corrosion process of copper materials can occur under some specific conditions, especially in aerated acidic media, which involves an enormous economic loss [15]. To resolve such problem, corrosion inhibitor was used as an efficient method to stop or reduce the degradation of copper materials [16].

---

[2] Before this year (i.e. the 1900s), used corrosion inhibitors are inorganic type.

In the aim to evaluate and quantify the prevention property of "green" inhibitors against the corrosion of copper materials, direct and indirect investigation methods were employed. Concerning direct methods are including weight loss, the volume of liberated H_2 gas (in acidic media) and temperature variation. Whereas indirect methods regroups several DC and AC electrochemical methods, especially potentiodynamic polarization measurements and electrochemical impedance spectroscopy. Furthermore, in recent works, some authors were used electrochemical frequency modulation (EFM). Due to their simplicity of performing, high precision, possibility to understand the action mechanism, low material and time consumption, electrochemical techniques are mostly used instead of direct methods.

In recent works, by using computational tools, corrosion scientists attempt to explain and understand more the action mechanism of those sustainable corrosion inhibitors under the atomic scale. Such interesting in using theoretical approaches related to the improvements of software and hardware performance of existing computers as well as the development of calculation algorithms in the last decade [17-22].

In the rest of this chapter, we will focus on the application of some biomolecules, namely amino acids and biopolymers, in addition to plant extracts and ionic liquids compounds as sustainable inhibitors toward the corrosion of copper and its based alloys in various corrosive media.

2.   Amino acids as sustainable inhibitors against the corrosion of copper materials

The protection capacity of numerous amino acids or their derived compounds to behave as eco-friendly inhibitors toward copper materials corrosion was exanimated and reported in the literature by several researchers.

Those biomolecules showed a specific significance to consider them as good environmentally inhibitors of corrosion. This explains by their interesting intrinsic properties such as biodegradability property and nontoxicity effect. In addition, it can be produced at a high level of purity and has a higher solubility in aqueous solutions and economically rentable [23-25].

The structure of an amino acid molecule is shown in Fig. 2 (in the center of Fig. 2). It characterized by the presence of amine and carboxyl functional groups. Almost tested amino acid compounds as corrosion inhibitors were the alpha (α) type, in which amine and carboxyl groups are bonded to the same carbon atom as demonstrated in Fig. 2. For the rest of the molecule skeleton, it can be an H atom (i.e. *Glycine,* it is the simplest amino acid) or an organic side chain of different sizes and shapes. Fig. 2 summarized the

Sustainable Corrosion Inhibitors   Materials Research Forum LLC
Materials Research Foundations**107** (2021) 175-203   https://doi.org/10.21741/9781644901496-8

molecules of amino acid present in physiological media, which contributed to proteins building *in vivo* species starting from cell to human [24].

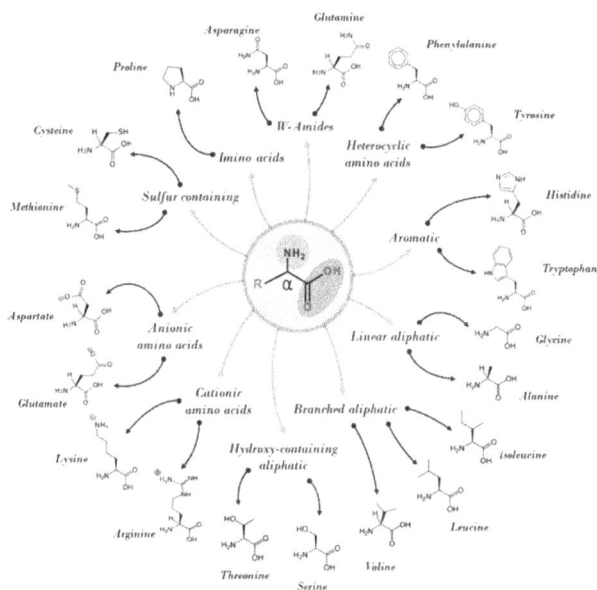

Figure 2.   *Molecular structure of amino acid and physiological amino acids [10].*

The amino acid compounds are used in numerous applications like in the food industry, animal feed technology, pharmaceuticals, cosmetics, and human parenteral nutrition. All these applications and other ones have involved the progress of amino acid-based compounds market. With respect to their environmental aspect, the existence of heteroatoms, like oxygen, nitrogen and sulfur atoms, in addition to eventual π-electrons on their molecular skeleton has encouraged several researchers to study in depth their possible performance to act as the sustainable inhibitors for corrosion of copper materials in different corrosive mediums [26-29]. Mainly for brass and copper/nickel alloys.

The use of amino acid-based compounds as inhibitors for copper or its alloys[3] was developed considerably throughout the latest years. In the first studies, the investigations were focused on the use of simple amino acids, i.e. physiological amino acids (Fig. 2). In

_____

[3] Mainly for CuZn and CuNi alloys.

this context, among tested biomolecules, *Cysteine* and *Methionine* have been studied extensively (as can be seen in Table 1), this particular attention is due to the presence of sulfur atoms, as well as nitrogen and oxygen atoms on their molecular structures, which can consist the actives sites for adsorption on copper surface [10]. Table 1 resumes some obtained protection effectiveness of simple amino acids for copper or its alloys.

Table 1.   Application of some simple amino acids as corrosion inhibitors toward the corrosion of copper materials.

Amino acids	Aggressive medium	Copper material	Inhibition efficiency (IE)	Ref
Aspartic acid, Glutamic acid, Asparagine and Glutamine	0.5 M HCl	Cu	IE(%): *Glutamine (73%) > Asparagine > Glutamic acid > Aspartic acid* at 10 mM.	[34]
Methionine, Cysteine, Serene, Arginine, Glutamine and Asparagine	0.1 M H_2SO_4	Cu	IE(%): *Arginine > Glutamine > Asparagine > Methionine > Cysteine > Serene* at 10 mM	[92]
Proline, Cysteine, Phenylalanine, Alanine, Histidine and Glycine	8 M H_3PO_4	Cu	*Histidine, Cysteine* and *Phenylalanine* give the maximum efficiency	[93]
Cysteine and Methionine	Neutral 3.5% NaCl	Cu10Al5Ni	IE(%): *Cysteine (96%) > Methionine (77%)* at 6 mM.	[94]
Cysteine, Alanine and Phenylalanine	0.2 g L^{-1} Na_2SO_4 + 0.2 g L^{-1} NaHCO$_3$ at pH 3	Bronze	IE(%): *Cysteine (97% at 10mM) > Alanine (80% at 1 M) > Phenylalanine (52% at 10 mM)*	[95]
Glycine, Alanine, Leucine, Cysteine, Lysine, Histidine and Glutamic acid	Neutral 0.06 M NaCl	CuNi5 and CuNi65	*Glycine* is an effective corrosion inhibitor	[96]

Despite that, the higher protection performance was judged by using high concentration levels of those amino acids. To overcome such deficiency and to enhance more their protection performance at lower concentrations, the corrosion scientists were carried out by applying the synergistic corrosion inhibiting effect. The synergism can be described as follow: the effect of combined substances is better as compared to the summation of separate actions of these substances [30,31]. In the field of corrosion inhibiting, the synergism effect can be reached by two different ways, the first one is by the interaction between components of the inhibitor formulation, whereas the second way is by the interaction of one component and the species present in the aqueous solution.

The synergistic corrosion inhibiting effect with simple amino acids was attained via the addition of different compounds. The effect of cations and anions species on the effectiveness of simple amino acids have been widely examined, as well as, the action of mixing two inhibitors (i.e. amino acid + organic compound). Among tested cations, zinc (Zn^{2+}) and cerium (Ce^{4+}) cations have received significant consideration [32,33], whereas, in the case of anions, they were the halide species [34-37].

For instance, Zhang *et al.* [32] stated that the copper corrosion in acidic media was reduced significantly using *Methionine* (12 mM) + Zn^{2+} (0.5 mM) formulation. The authors explain that by the formation of *Methionine* – Zn^{2+} complex, which forms upon the copper surface a protective layer. In another study [33], utilizing PDP and EIS techniques and weight loss measurements, the formulation consisting of *L-phenylalanine* at 5 mM and Ce^{4+} at 2 mM was found as an effective inhibitor against copper corrosion in hydrochloride medium. In the first stage, *L-phenylalanine* had shown a limited inhibitive action, but by adding Ce^{4+} species in the second stage, it was enhanced. This effect was attributed to the formation of adsorbed *L-phenylalanine* – Ce^{4+} complex on the copper surface.

Concerning the synergistic effect caused by halide species, especially iodide anion, it has been reported that in the presence of this anion the prevention effectiveness of amino acids was improved. Nazeer *et al.* [36] have been reported an interesting synergistic corrosion inhibiting effect of iodide ions with *Glycine* for copper-nickel alloy immersed in sulfide polluted saline solution, which at 500 ppm of *Glycine* and 100 ppm of iodide anion, the obtained prevention efficiency reached 87.4 %. In acidic medium, the addition of KI was also shown a benefic effect on the protection performance of *Glutamine* for copper [34].

On another side, many investigations have been performed to examine the influence of amino acids/organic compounds mixtures on the corrosion inhibition process. Those organic mixtures were consisting of a melange of amino acid compounds themselves or amino acid with other organic compounds. For example, it has been reported [38] that *Glycine/Glutamic acid/Cysteine* amino acid mixture can be used as a potent corrosion inhibitor for copper in HCl medium. In similar aggressive media, it has been found [39] that *Methionine*/Cetrimonium bromide combination exhibits a good performance to reduce copper dissolution. These studies were performed employing PDP, EIS and cyclic voltammetry techniques.

Consequently, the synergism effect can be considered as a useful process to enhance the inhibition potency of simple amino acids, to decrease used concentrations and to expand the application of these biomolecules for copper materials in different aggressive environments.

The researches on using amino acid compounds as eco-friendly inhibitors against copper materials dissolution were not limited to employing known physiological amino acids or by applying the synergistic method, but it was extended to use amino acid derivatives. For this purpose, the new challenge for corrosion scientists was the synthesizing of new effective eco-friendly amino acid derivatives. According to corrosion literature, some derivatives were synthesized and exanimated as potential effective inhibitors to decline the copper materials corrosion in various environments. For instance, Khaled [40] have examined the inhibitive effect of *Methionine* and two corresponding derivatives, namely methionine sulfone and methionine sulfoxide (Fig. 3), on the dissolution rate of copper in 1 M HNO_3 solution. According to experimental findings, the inhibition performance increases as follow: *Methionine* sulfone (90.7 %) > *Methionine* sulfoxide > *Methionine* (81.6 %) at 5 mM. The observed comportment had been related to their structural parameters applying DFT and molecular dynamics calculations.

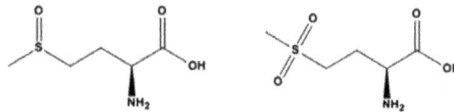

Figure 3.   Molecular structures of L-methionine sulfoxide (left) and L-methionine sulfone (right).

On the subject of the application of amino acid-based derivatives as inhibitors toward copper materials corrosion, the number of carried studies are limited as compared to theirs in the case of iron and its alloys, which are received more attention by corrosion scientists. Such trend can be related to tremendous industrial applications of iron *Vis* copper materials [10].

3.   Biopolymers as sustainable inhibitors against the corrosion of copper materials

Biopolymers products are commonly used as metal protective coatings. Due to their zero pollution impact, biodegradability and chemical stability, those compounds meet the requirements to be applied as efficient ecological inhibitors of corrosion for copper materials.

Those bio-macromolecules are extracted from natural sources (flora or fauna), which considered as renewable and low-cost alternatives of widely used corrosion inhibitors in industrial applications. Numerous carbohydrate macromolecules are reported in corrosion literature as potent inhibitors against metals dissolution in different media, such as chitosan, alginate, hydroxyethyl/carboxymethyl cellulose, starch, exudate gums, pectin, pectates, carrageenan, and dextrin/cyclodextrins biopolymers. Fig. 4 illustrates the molecular structures of some investigated biopolymers as corrosion inhibitors. Concerning their protective action, in most cases, it was attributed to the biopolymer chemical composition (i.e. type and number of presented heteroatoms) and their molecular weights, further to their electronic and molecular structures [41].

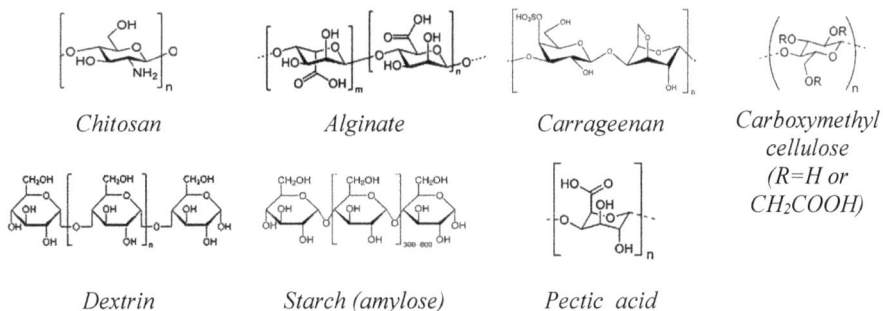

| Chitosan | Alginate | Carrageenan | Carboxymethyl cellulose (R=H or CH₂COOH) |

| Dextrin | Starch (amylose) | Pectic acid |

Figure 4.   Molecular structure of some studied bio-macromolecules monomers

Among the above-listed biopolymers, chitosan and alginate biopolymers have been investigated for copper materials immersed in several aggressive environments [42-45], which showed a higher performance to reduce the corrosion rate of these materials. Chitosan is extracted mainly from two marine crustaceans, namely shrimp and crabs. This biopolymer is used in many applications, such as agriculture (e.g. fertilizer), food, packaging (i.e. bioplastic), water treatments and medical domains. It is well known that chitosan has a good complexing ability related to –NH₂ groups. Whereas, alginate is extensively extracted from brown seaweed, like *Fucus* or *Laminaria*, which commonly used in textures, drink and food products [46].

For instance, in molar hydrochloric acid solution, chitosan was exhibited 86% as inhibition efficacy for copper at 0.1 mg L^{-1} [42]. The protective action of investigated bio-macromolecule was confirmed through visual observation as well as via SEM (Fig. 5) and AFM images. In recent work, El Mouaden *et al.* [43] have reported similar behaviors of chitosan in a sulfide-containing synthetic seawater environment, in which the inhibition effectiveness reached 89% at 800 ppm for copper. According to these studies, even at lower concentrations and higher temperatures levels as well as for long immersion times (from 7 to 90 days), chitosan was demonstrated a remarkable protection action against copper corrosion.

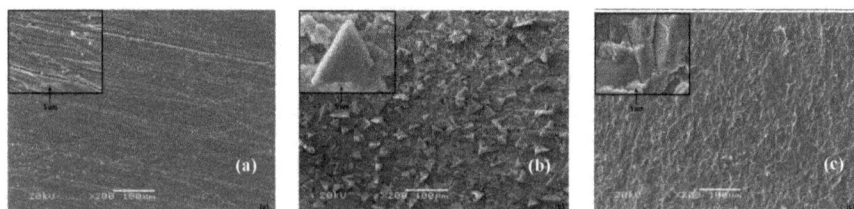

*Figure 5.   SEM images of polished copper sample before (**a**) and after its 24 h immersion (**b**) without and (**c**) with chitosan in molar HCl solution. Reprinted by permission from Springer [42] [4740431036855].*

In saline solution (3 wt% NaCl), Oukhrib *et al.* [44] have been confirmed the protective action of chitosan against copper degradation that is 87% at 1 gL^{-1}. It is known that in solution, macromolecules can exist with different chain lengths (i.e. fragments). Under this consideration, the authors have been carried out theoretical studies on the interaction of different chitosan fragments with the copper surface. Based on *Monte Carlo* simulations, they were found that the length of the chitosan biopolymer was affected noticeably its mode of adsorption into the copper surface, indeed its adsorption energy. This shows the importance of employing theoretical tools to define and clarify the inhibition corrosion mechanism of those bio-macromolecules.

The obtained inhibition effect was attributed to the adsorption of chitosan fragments into copper surface via highest negative charges centers located on nitrogen and oxygen atoms or by complexing metal ions on the surface via lone pair electrons of amine and hydroxyl groups, which involves the establishment of protective layer upon the metallic surface.

These works confirmed the possibility to apply chitosan biopolymer as an effective sustainable anti-corrosion compound for copper materials under various conditions.

Concerning alginate biopolymer, it was exhibited a good performance against copper corrosion in acidic medium. According to Jmiai *et al.* [45], a maximum efficiency of 83% was achieved at a lower concentration of alginate (0.1 mg L^{-1}). The researchers have been reported that obtained experimental data are in good agreement with computational ones.

As it has been reported for amino acid-based compounds, studies were extended to synthesize new derivatives. Such trend attributes to the interest of corrosion scientists to improve more the anti-corrosion performance of exist biomolecules or to make them more soluble in aqueous media. For this purpose, several biopolymer derivatives were synthesized and tested as inhibitors of corrosion for numerous metals, especially iron-based metallic materials. However, the studies on copper materials are limited and need to be developed in future works. In the available literature, there is an alone study on copper conducted by Kumar *et al.* [47] in aqueous chloride solution. In this work, the authors have modified chemically an existing biopolymer (i.e. chitin) to produce a new derivative, namely phosphorylated chitin (PCT, Fig. 6). Maximum effectiveness of 92 % was attained by the addition of 200 ppm of modified biopolymer. The reduction of copper dissolution rate was attributed to the formation of a protective film involving Cu^{2+} – PCT complex.

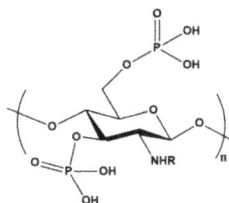

Figure 6.   Molecular structure of PCT monomer (where R=H or COCH$_3$ and n(–COCH$_3$) > n(–H)).

The possibility to rise the inhibition efficiency of biopolymers via synergistic effect has not yet been explored for copper materials, which consists a new research axis that needs to be developed in future works. The same attention must be devoted to the ability of other biopolymers to act as sustainable inhibitors for copper-based metallic materials, e.g. hydroxyethyl/carboxymethyl cellulose, starch, exudate gums, pectin, pectate, carrageenan, and dextrin/cyclodextrins carbohydrate polymers. On the other hand, the application of biopolymers, or in general organic compounds with high molecular weight, can be limited by their low solubility in the aqueous aggressive solutions [48]. For instance, poor solubility can overcome by chemical modification on the molecular structure of those compounds [49].

4.   Natural extracts as sustainable inhibitors against the corrosion of copper materials

To face the degradation of copper materials and to replace environmentally unfriendly corrosion inhibitors compounds, the corrosion scientists reported the benefic action of natural origin compounds, which can be promising green anti-corrosion compounds. The trend to employ natural outcome substances as protective agents explained by their biodegradability, renewability and ecologically acceptability as well as economically advantageous [50].

Some research groups [51,52] have reported the good prevention property of some plant oils, like *Argan* oils. However, such trend is limited by the lower solubility of tested oils in used media, which involves the addition of supplementary organic solvents. Another deficiency that faces against the use of oils as sustainable inhibitors is not economically profitable. These weaknesses encourage more the evaluation of various plant extracts as sustainable corrosion inhibitors. The great interest in these natural extracts in recent years, also, due to the ability of most metabolites existing in the plant (i.e. flavonoids, catechins, organic acids, phenolic compounds, quinone, alkaloids, terpenoids) to be soluble in aqueous media [53]. Consequently, the obtained extract will contain numerous organic compounds with heteroatoms, aromatic heterocyclic and aliphatic rings, which can serve as actives adsorption sites. It is important to mention that the composition of these natural mixtures depends on several parameters like season, harvesting location and weather conditions.

Concerning, the prevention effect of an extract, it can be attained by the action of a specific extract component (i.e. molecule) or by a synergistic effect of many components. This (these) molecule(s) adsorbs upon the metal surface or complex metallic cations to form protecting organic layer, which separates metal surface from corrosive agents existing in aggressive environment [54].

On the other hand, the extraction effectiveness of a plant extract depends on many factors, mainly to used solvent for the extraction process and the fixed extraction temperature [55,56]. The solvent allows extracting the compounds present in plant materials by its diffusion into their tissues and then solubilized them. The available literature [57,58] reveals that water is widely used to perform the extraction process as compared to ethanol and methanol. This is due to its non-toxic nature, simplicity to use, inflammable and readily availability. Relating to temperature, generally, the extract preparation is performed in 60-80 °C temperature range; because at lower levels, it limits the solubility of extract compounds, whereas, at higher levels, it can cause decomposition of these constituents.

The extracts of different parts of the same plant (e.g. seeds, leaves, root, and stem) were tested and reported to act as the green corrosion inhibitors in various electrolytic media. Generally, the leave extract was exhibited remarkable anti-corrosion activity as compared to other plant parts extracts. The noted difference in prevention activities is related to existing of different chemical components in extracts of each plant parts, in which, leaves are richer ones (due to photosynthesis process in leaves) [59].

According to corrosion literature, the researches concerning this topic has exponentially amplified over the past decade. In the rest of this section, we will cite the major findings of some collected works corresponding to the use of natural extracts as sustainable inhibitors for copper materials.

It has been reported that *vitex negundo* leaves extract can be served as an effective green anti-corrosion compound in 3 M nitric acid solution [60]. The electrochemical measurements confirm an inhibition percentage of over 90% using 0.1 gL^{-1} of extract at 35 °C for four hours of immersion. In another investigation carried out on extract of leaves, Rahal *et al.* [61] underlined the benefic effect of *olive* leaves extract on the copper degradation in slain medium, in which the inhibition efficiency reached 86% at 2.42 mM of extract.

Serving the synergistic corrosion inhibiting effect, Fouda *et al.* [62] claimed a good inhibition effect of "aqueous extract of *lupine* seed + barium chloride" formulation in acidic medium for copper, which the inhibition effectiveness attained 90% (at 250 ppm + 10^{-3} M). The synergistic effect can be carried out also by combing two extracts as reported by Loto *et al.* [63]. In this work, the mixture of *Carica Papaya* and *Camellia Sinensis* leaves extracts was exhibited appreciable performance to control brass[4] corrosion in molar HNO$_3$ solution.

The application of green extract was not focused on acidic and near-neutral media, but it was extended to alkaline solutions [64,65]. In this context, we cite the work carried out by Wedian *et al.* [65] on *Capparis spinose* extract in molar hydroxide aqueous solution. Under these conditions, a maximum of protection activity of 74% was obtained at 400 ppm of extract.

---

[4] $^{65}Cu^{35}Zn$ alloy.

Materials Research Foundations 107 (2021) 175-203   https://doi.org/10.21741/9781644901496-8

In addition to all listed strongest features on using plant extract for copper protection, some extracts showed an excellent ability to be applied in different aggressive environments [64-68]. For example, the aqueous extract of *Jujube* (*Ziziphus lotus*) was exhibited an interesting reduction effect of copper dissolution in different media, namely acidic solution (93% at 1 gL^{-1}) [67] and natural seawater (83% at 1 gL^{-1}, see Fig. 7) [66]. Table 2 collects the inhibition effectiveness, IE(%), of some other plant extracts employed as sustainable anti-corrosion compounds for copper in different media.

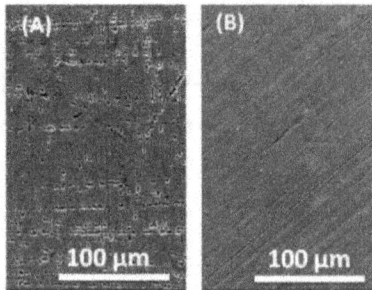

*Figure 7.   SEM images of the copper surface after it immersion in natural seawater (**A**) without and (**B**) with aqueous extract of Jujube in natural seawater of Agadir, Morocco [66].*

Table 2.   IE(%) of some natural extracts employed as corrosion inhibitors for copper materials

Plant extract	Metal	Aggressive medium	Extract concentration	IE(%)	Ref
Hyoscyamus Muticus	Cu	1 M HNO$_3$	0.5 g L^{-1}	89	[97]
Funugreek	Cu	2 M H$_2$SO$_4$	100 mL L^{-1}	84	[98]
Euphorbia helioscopia linn	Cu	1 M HNO$_3$	0.5 g L^{-1}	91	[99]
Capparis decidua	Cu	0.5 M HCl	≈ 1 g L^{-1}	51	[76]
Egyptian licorice	Cu	0.1 M HCl	8% (v/v)	90	[100]
Veronica rosea	Cu	1 M HNO$_3$	0.3 g L^{-1}	87	[74]
Myrrh	Cu-33Zn alloy	3.5% NaCl + 16 ppm S^{-2}	0.3 g L^{-1}	67	[101]
Emblica officinalis	Cu and Cu-27Zn alloy	Natural seawater	1 g L^{-1}	80	[102]

In recent years, the theoretical approach has been widely employed to discuss and understand the action mechanism of corrosion inhibitors under the atomic scale, as well as to support obtained experimental results [19,69-71]. In the case of natural extract inhibitors, such depth approach was infrequently used for copper [72]. Taking it in the account and in the absence of further data support, probable interpretations (hypothesis) of observed protection behaviors were suggested. For instance, without a strong background and uncomplete vision on the studied inhibition system, some authors attributed the corrosion prevention of plant extract to complexes formation, whereas other ones to the adsorption mechanism [73]. In term of constituents, almost tested plant extracts remain "unidentified mixtures" that explains the origin of miss-understanding of the inhibition process. Despite that, recently, some efforts to characterize the main constituents of plant extracts, tested as corrosion inhibitors, were highlighted in the literature [62,74-77]. The main composition of some investigated green extracts for copper materials is shown in Table 3.

Table 3.   Main constituents present in some plant extracts used as sustainable corrosion inhibitors for copper materials

Plant extract	Major constituents	Ref
Euphorbia heterophylla	Sterols; flavonoids; triterpenoids; saponins; alkaloids and tannins	[62]
Propolis	Luteolin; Prenylated coumarin; 7-O-prenylpinocembrin; cearoin; P-methoxycinnamic acid; 6-methoxy deriphyllin; linalool; 3-phenyl resveratrol and 3-prenyl-4-hydroxy-suberosin cinnamic acid	[75]
Capparis decidua seeds	Gluco-capparin; L-Stachydrine; β-sitosterol; L-3-Hydroxy stachydrine	[76]
Garlic	1-Oxa-4,6-diazacyclooctane-5-thione; 1,3-Dithiane; Propionic acid; Di-2-propenyl disulfide; 2-Methyltetrahydrothiophene; Diallyl trisulfide and 2-Prop-2-enylsulfanylacetonitrile	[77]

5.   Ionic liquids as sustainable inhibitors against the corrosion of copper materials

By definition, ionic liquids are principally composed of ion species with a lower melting point (< 100 °C). Following the required purpose, characteristics of these liquids can be changed by selecting the appropriate anion and cation species. Ionic liquids can be categorized into different groups. According to Hajipour and Refiee [78], there are 11 classes of ionic liquids, e.g. basic, acid, protic, supported and poly-ionic liquids. Another classification has been reported by Suresh and Sandhu [79], in which ionic liquids are subdivided into two major classes, namely anionic and cationic ionic liquids. Table 4 collects the molecular structure of some ionic liquids compounds belong to basic and acidic classes (according to Hajipour's and Refiee's classification).

Table 4.   *Molecular structure of some ionic liquids compounds.*

Basic class	
	(Common anions: Cl^-, CH_3OO^- ...)
Acidic class	
	(Common anions: $H_2PO_4^-$, HSO_4^- ...)

Ionic liquids are characterized by several attractive properties such as their non-inflammability, negligible volatility, high thermal and chemical stability, as well as their high ability to dissolve almost organic and inorganic compounds and longtime stability. Furthermore, ionic liquids can be dissolved at a high level in aqueous solutions [80]. Taking into account the "green chemistry" aspect, ionic liquids are classified as green compounds due to their zero or little negative impact on health and environment [81].

All listed characteristics of this class of compounds make them potential candidates to substitute conventional organic solvents. This involves its utilization in many applications including solvent extraction, solvent for polymerization, synthesis of nanomaterials, separation of petrochemical relevance, chemical analysis, catalyst of chemical and biochemical transformations [81-83].

In recent years, the evaluation of ionic liquids as sustainable corrosion inhibitors to replace conventional inhibitors was the subject of many works and being an active topic around the "green corrosion inhibitors".

Agreeing to the literature [84], ionic liquids were largely reported as efficient corrosion inhibitors for iron-based metallic materials as well as non-ferrous ones immersed in different aggressive media. The use of these liquids on ferrous metals or its alloys were received more attention from corrosion scientists as compared to non-ferrous ones, like copper and its alloys. Among existing ionic liquids, imidazole derivatives have been extensively employed [85,86].

Similar to tradition inhibitors, the anti-corrosion effect of ionic liquids is attributed to the formation of protecting film upon the surface of metal via their direct or indirect (via preadsorbed species on the metal surface, like anions) adsorption. This process involves the isolation of metal from the corrosives species present in media.

In order to describe and understand their adsorption behaviors on the metallic substrate, generally, DFT/B3LYP[5] computations and molecular mechanic simulations have been used. In most cases, a good agreement was outlined between observed inhibition effectiveness and the majority of obtained computational data (e.g. molecular dipole moment, HOMO[6], LUMO[7], ΔE and E$_{ads}$ energies) [87-89].

At ambient temperature, Feng *et al.* [87] have been reported the excellent anti-corrosion action of thiazolyl blue (Fig. 8 (a)) ionic liquid for the copper in chloride solution. An efficiency of 92% was attained at 5 mM of tested organic inhibitor. Based on spectral (i.e. XPS and FT-IR) and contact angle techniques, this effect was explained by the formation of protective hydrophobic film (Fig. 9) on the metal surface by the adsorption of formed thiazolyl blue – metal complex (via nitrogen and sulfur atoms). The effective protection of tested ionic liquids against copper degradation is clear in Fig. 9.

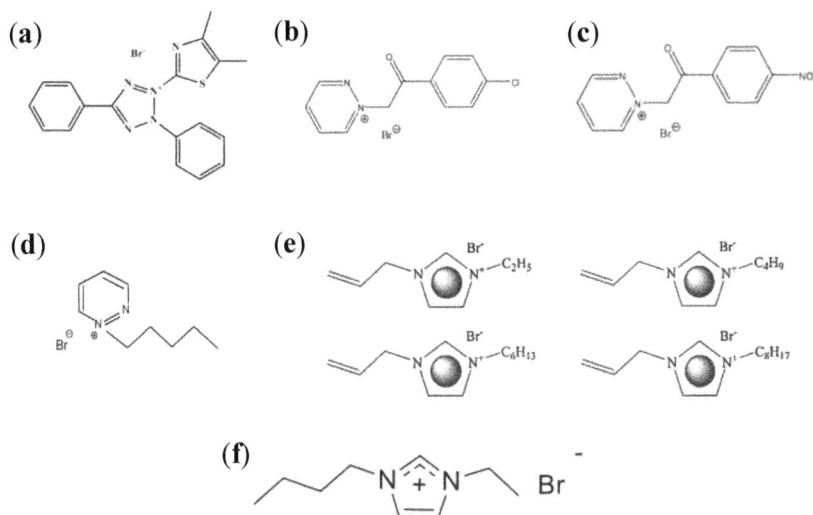

Figure 8.   *Molecular structure of some tested ionic liquids for copper*

---

[5] Density Functional Theory at Lee–Yang–Parr correlation functional.
[6] Highest Occupied Molecular Orbital.
[7] Lowest Unoccupied Molecular Orbital.

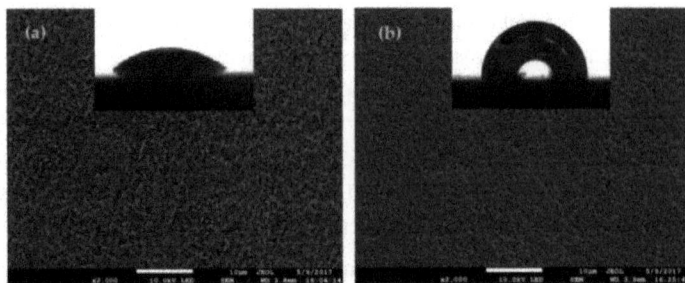

Figure 9.   SEM images and contact angle of the copper surface immersed in simulated seawater (a) without and (b) with thiazolyl blue ionic liquids [87].

The application of ionic liquids as corrosion inhibitor for copper was not limited to existing ones, but it was prolonged to develop new synthesized derivatives. For instance, in a bi-molar phosphoric acid solution containing chloride anions, Bousskri *et al.* [88, 89] revealed the good ability of three new synthetized Pyridazinium derivatives to act as eco-friendly inhibitors for copper corrosion. The evaluated compounds are [1–(2–(4–chlorophenyl)–2–oxoethyl)– (Fig. 8 (b), **CPEPB**), [1–(2–(4–nitrophenyl)–2–oxoethyl)– (Fig. 8 (c), **NPEPB**), and 1–pentyl–pyridazinium bromid (Fig. 8 (d), **PPB**) derivatives. The maximum effectiveness of 87, 88 and 93% were achieved at 1 mM of CPEPB, NPEPB and PPB, respectively. The authors affirmed the good agreement between obtained experimental and theoretical results.

Likewise, Qiang *et al.* [90] have been synthesized four allyl imidazolium compounds with different chain lengths (Fig. 8 (e)) and used them as inhibitor compounds against the corrosion of copper in 0.5 M H_2SO_4 medium. In the aim to investigate that, the authors were combined experimental and computational approaches. The developed molecular liquids were exhibited unfavorable action toward the dissolution rate of copper, which ascribed to the formation of protective adsorbed molecular layer of tested compounds on the copper surface. It was reported that the prevention effectiveness increases by raising chain length, which reaches 98% at 1 mM for the fourth allyl imidazolium derivative (i.e. with $-C_8H_{17}$). Such protection level shows the ability of ionic liquids to replace traditional toxic corrosion inhibitors (e.g. chromate, some organic inhibitors) to decline copper degradation in acidic media.

Similar work as previous ones, new ionic liquid-based compound was synthesized, namely 1-butyl-3-ethylimidazolium bromide (Fig. 8 (f)), then tested for copper

Materials Research Forum LLC
https://doi.org/10.21741/9781644901496-8

dissolution prevention in 0.1 M Na_2SO_4 (at pH = 3) [91]. According to obtained experimental data, high inhibition performance (over 90%) is recorded at low concentration (i.e. 1 mM). This reduction of the corrosion process was related to the isolation of almost copper surface from corrosives agents via a protecting film. The latest film was formed by the pre-adsorption of bromide ions on the copper surface followed by the interaction of carbon atoms of the tested molecule with it.

References

[1]   V.S. Sastri, E.Ghali, M. Elboujdaini, Corrosion Prevention and Protection Practical Solutions, first ed., John Wiley & Sons Ltd, 2007. https://doi.org/10.1002/9780470024546

[2]   G.H. Koch, N.G. Thompson, O. Moghissi, J.H. Payer, J. Varney, IMPACT (International Measures of Prevention, Application, and Economics of Corrosion Technologies Study), NACE International, Houston, TX, 2016.

[3]   A.A. Olajire, Corrosion inhibition of offshore oil and gas production facilities using organic compound inhibitors - A review, J. Mol. Liq. 248 (2017) 775–808. https://doi.org/10.1016/j.molliq.2017.10.097

[4]   Y. Kharbach, F.Z. Qachchachi, A. Haoudi, M. Tourabi, A. Zarrouk, C. Jama, L.O. Olasunkanmi, E.E. Ebenso, F. Bentiss, Anticorrosion performance of three newly synthesized isatin derivatives on carbon steel in hydrochloric acid pickling environment: Electrochemical, surface and theoretical studies, J. Mol. Liq. 246 (2017) 302–316. https://doi.org/10.1016/j.molliq.2017.09.057

[5]   E. Ech-chihbi, M.E. Belghiti, R. Salim, H. Oudda, M. Taleb, N. Benchat, B. Hammouti, F. El-Hajjaji, Experimental and computational studies on the inhibition performance of the organic compound "2-phenylimidazo [1,2-a]pyrimidine-3-carbaldehyde" against the corrosion of carbon steel in 1.0 M HCl solution, Surf. and Interfaces. 9 (2017) 206–217. https://doi.org/10.1016/j.surfin.2017.09.012

[6]   I. B. Obot, M. M. Solomon, S. A. Umoren, R. Suleiman, M. Elanany, N. M. Alanazi, A. A. Sorour, Progress in the development of sour corrosion inhibitors: Past, present, and future perspectives, Journal of Industrial and Engineering Chemistry 79 (2019) 1–18. https://doi.org/10.1016/j.jiec.2019.06.046

[7]   V.S. Sastri, Green corrosion inhibitors, Theory and Practice, John Wiley & Sons, Inc., New Jersey, 2011. https://doi.org/10.1002/9781118015438

[8] P.B. Raja, M.G. Sethuraman, Natural products as corrosion inhibitor for metals in corrosive media - A review, Mater. Lett. 62 (2008) 113–116. https://doi.org/10.1016/j.matlet.2007.04.079

[9] V.S. Sastri, Green Corrosion Inhibitors: Theory and Practice, first ed., John Wiley & Sons Ltd, 2011. https://doi.org/10.1002/9781118015438

[10] B. El Ibrahimi, A. Jmiai, L. Bazzi, S. El Issami, Amino acids and their derivatives as corrosion inhibitors for metals and alloys, Arab. J. Chem. 13 (2020) 740-771. https://doi.org/10.1016/j.arabjc.2017.07.013

[11] P. Zhang, Q. Zhu, Q. Su, B. Guo, S.-k. Cheng, Corrosion behavior of T2 copper in 3.5% sodium chloride solution treated by rotating electromagnetic field, Trans. Nonferrous Met. Soc. China. 26 (2016) 1439-1446. https://doi.org/10.1016/S1003-6326(16)64249-8

[12] L. Núñez, E. Reguera, F. Corvo, E. González, C. Vazquez, Corrosion of copper in seawater and its aerosols in a tropical island, Corros. Sci. 47 (2005) 461-484. https://doi.org/10.1016/j.corsci.2004.05.015

[13] S. Suzuki, N. Shibutani, K. Mimura, M. Isshiki, Y. Waseda, Improvement in strength and electrical conductivity of Cu–Ni–Si alloys by aging and cold rolling, J. Alloys Compd. 417 (2006) 116-120. https://doi.org/10.1016/j.jallcom.2005.09.037

[14] H.-x. Wang, Y. Zhang, J.-l. Cheng, Y.-s. Li, High temperature oxidation resistance and microstructure change of aluminized coating on copper substrate, Trans. Nonferrous Met. Soc. China. 25 (2015) 184-190. https://doi.org/10.1016/S1003-6326(15)63594-4

[15] A.H. Moreira, A.V. Benedetti, P.L. Calot, P.T.A. Sumodjo, Electrochemical behaviour of copper electrode in concentrated sulfuric acid solutions, Electrochim. Acta. 38 (1993) 981–987. https://doi.org/10.1016/0013-4686(93)87018-9

[16] M.M. Antonijevic, M.B. Petrovic, Copper corrosion inhibitors. A review, Int. J. Electrochem. Sci. 3 (2008) 1–28.

[17] G. Gece, The use of quantum chemical methods in corrosion inhibitor studies, Corros. Sci. 50 (2008) 2981-2992. https://doi.org/10.1016/j.corsci.2008.08.043

[18] S. Kaya, L. Guo, C. Kaya, B. Tüzün, I.B. Obot, R. Touir, N. Islam, Quantum chemical and molecular dynamic simulation studies for the prediction of inhibition efficiencies of some piperidine derivatives on the corrosion of iron, J. Taiwan Inst. Chem. Eng. 65 (2016) 522-529. https://doi.org/10.1016/j.jtice.2016.05.034

[19]   B. El Ibrahimi, A. Soumoue, A. Jmiai, H. Bourzi, R. Oukhrib, K. El Mouaden, S. El Issami, L. Bazzi, Computational study of some triazole derivatives (un- and protonated forms) and their copper complexes in corrosion inhibition process, J. Mol. Struct. 1125 (2016) 93-102. https://doi.org/10.1016/j.molstruc.2016.06.057

[20]   M. K. Awad, M. R. Mustafa, M.M.A. Elnga, Computational simulation of the molecular structure of some triazoles as inhibitors for the corrosion of metal surface, J. Mol. Struct. THEOCHEM. 959 (2010) 66-74. https://doi.org/10.1016/j.theochem.2010.08.008

[21]   A. Aouniti, K. F. Khaled, B. Hammouti, Correlation Between Inhibition Efficiency and Chemical Structure of Some Amino Acids on the Corrosion of Armco Iron in Molar HCl, Int. J. Electrochem. Sci. 8 (2013) 5925-5943.

[22]   N.O. Eddy, U.J. Ibok, B.I. Ita, QSAR and quantum chemical studies on the inhibition potentials of some amino acids for the corrosion of mild steel in H_2SO_4, J. Comput. Methods Sci. Eng. 11 (2011) 25-43. https://doi.org/10.3233/JCM-2011-0290

[23]   D.Q. Zhang, B. Xie, L.X. Gao, Q.R. Cai, H.G. Joo, K.Y. Lee, Intramolecular synergistic effect of Glutamic acid, Cysteine and Glycine against copper corrosion in hydrochloric acid solution, Thin Solid Films. 520 (2011) 356–361. https://doi.org/10.1016/j.tsf.2011.07.009

[24]   M. S. Kilberg, D. Häussinger, Mammalian Amino Acid Transport, Springer Science & Business Media, 1992. https://doi.org/10.1007/978-1-4899-1161-2

[25]   B. El Ibrahimi, A. Jmiai, A. Somoue, R. Oukhrib, M. Chadili, S. El Issami, L. Bazzi, Cysteine Duality Effect on the Corrosion Inhibitionand Acceleration of 3003 Aluminium Alloy in a 2% NaCl Solution, Port. Electrochimica Acta. 36(6) (2018) 403-422. https://doi.org/10.4152/pea.201806403

[26]   V.F. Wendisch, Microbial production of amino acids and derived chemicals: Synthetic biology approaches to strain development, Curr. Opin. Biotechnol. 30 (2014) 51-58. https://doi.org/10.1016/j.copbio.2014.05.004

[27]   J.-H. Lee, V.F. Wendisch, Production of amino acids – Genetic and metabolic engineering approaches, Bioresour. Technol. 245(B) (2017) 1575-1587. https://doi.org/10.1016/j.biortech.2017.05.065

[28]   S. Al-Dahir, N. Vithlani, A. Smith, J.F. Davis, S. Sirohi, Side Effects of Drugs Annual, Elsevier, 2017.

[29] G. Deferrari, I. Mannucci, G. Garibotto, Amino Acid Biosynthesis, in: J.W. Sons (Ed.), Encyclopedia of Life Sciences, 2010. https://doi.org/10.1002/9780470015902.a0000628.pub2

[30] S.A. Umoren, M.M. Solomon, Synergistic corrosion inhibition effect of metal cations and mixtures of organic compounds: A Review, J. Environ. Chem. Eng. 5 (2017) 246-273. https://doi.org/10.1016/j.jece.2016.12.001

[31] S.A. Umoren, M.M. Solomon, Effect of halide ions on the corrosion inhibition efficiency of different organic species – A review, J. Ind. Eng. Chem. 21 (2015) 81–100. https://doi.org/10.1016/j.jiec.2014.09.033

[32] D.Q. Zhang, Q.R. Cai, X.M. He, L.X. Gao, G.S. Kim, Corrosion inhibition and adsorption behavior of methionine on copper in HCl and synergistic effect of zinc ions, Mater. Chem. Phys. 114 (2009) 612–617. https://doi.org/10.1016/j.matchemphys.2008.10.007

[33] D.Q. Zhang, H. Wu, L.X. Gao, Synergistic inhibition effect of L-phenylalanine and rare earth Ce(IV) ion on the corrosion of copper in hydrochloric acid solution, Mater. Chem. Phys. 133 (2012) 981–986. https://doi.org/10.1016/j.matchemphys.2012.02.001

[34] D.Q. Zhang, Q.R. Cai, X.M. He, L.X. Gao, G.D. Zhou, Inhibition effect of some amino acids on copper corrosion in HCl solution, Mater. Chem. Phys. 112 (2008) 353–358. https://doi.org/10.1016/j.matchemphys.2008.05.060

[35] A.S. Fouda, A.A. Nazeer, E.A. Ashour, Amino acids as environmentally-friendly corrosion inhibitors for Cu10Ni alloy in sulfide-polluted salt water: Experimental and theoretical study, Mater. Prot. 52 (2011) 21-34.

[36] A. A. Nazeer, K. Y. Nageh, I. A. Gehan, A. Elsayed, Effect of Glycine on the electrochemical and stress corrosion cracking behavior of Cu10Ni alloy in sulfide polluted salt water, Ind. Eng. Chem. Res. 50 (2011) 8796-8802. https://doi.org/10.1021/ie200763b

[37] D. Zhang, X. He, Q. Cai, L. Gao, G. Kim, Arginine self-assembled monolayers against copper corrosion and synergistic effect of iodide ion, J. Appl. Electrochem. 39 (2009) 1193-1198. https://doi.org/10.1007/s10800-009-9784-7

[38] D.Q. Zhang, B. Xie, L.X. Gao, Q.R. Cai, H.G. Joo, K.Y. Lee, Intramolecular synergistic effect of glutamic acid, cysteine and glycine against copper corrosion in hydrochloric acid solution, Thin Solid Films. 520 (2011). https://doi.org/10.1016/j.tsf.2011.07.009

[39] D.Q. Zhang, B. Xie, L.X. Gao, H.G. Joo, K.Y. Lee, Inhibition of copper corrosion in acidic chloride solution by Methionine combined with cetrimonium bromide/cetylpyridinium bromide, J. Appl. Electrochem. 41 (2011) 491–498. https://doi.org/10.1007/s10800-011-0259-2

[40] K.F. Khaled, Corrosion control of copper in nitric acid solutions using some amino acids – A combined experimental and theoretical study, Corros. Sci. 52 (2010) 3225–3234. https://doi.org/10.1016/j.corsci.2010.05.039

[41] S.A. Umoren, U.M. Eduok, Application of carbohydrate polymers as corrosion inhibitors for metal substrates in different media: A review, Carbohyd. Polym. 140 (2016) 314–341. https://doi.org/10.1016/j.carbpol.2015.12.038

[42] A. Jmiai, B.E. Ibrahimi, A. Tara, R. Oukhrib, S.E. Issami, O. Jbara, L. Bazzi, M. Hilali, Chitosan as an eco-friendly inhibitor for copper corrosion in acidic medium: protocol and characterization, Cellulose. 24 (2017) 3843-3867. https://doi.org/10.1007/s10570-017-1381-z

[43] K.E. Mouaden, B.E. Ibrahimi, R. Oukhrib, L. Bazzi, B. Hammouti, O. Jbara, A. Tara, D.S. Chauhan, M.A. Quraishi, Chitosan polymer as a green corrosion inhibitor for copper in sulfide-containing synthetic seawater, Int. J. Biol. Macromol. 119 (2018) 1311–1323. https://doi.org/10.1016/j.ijbiomac.2018.07.182

[44] R. Oukhrib, B.E. Ibrahimi, H. Bourzi, K.E. Mouaden, A. Jmiai, S.E. Issami, L. Bammou, L. Bazzi, Quantum chemical calculations and corrosion inhibition efficiency of biopolymer "chitosan" on copper surface in 3%NaCl, J. Mater. Environ. Sci. 8 (2017) 195-208.

[45] A. Jmiai, B. El Ibrahimi, A. Tara, S. El Issami, O. Jbara, L. Bazzi, Alginate biopolymer as green corrosion inhibitor for copper in 1 M hydrochloric acid: Experimental and theoretical approaches, J. Mol. Struct. 1157 (2018) 408-417. https://doi.org/10.1016/j.molstruc.2017.12.060

[46] Y. Qin, J. Jiang, L. Zhao, J. Zhang, F. Wang, in: A.M. Grumezescu, A.M. Holban (Eds.), Biopolymers for Food Design, Academic Press, 2018, pp. 409-429. https://doi.org/10.1016/B978-0-12-811449-0.00013-X

[47] K.V. Kumar, B.V.A. Rao, N.Y. Hebalkar, Phosphorylated chitin as a chemically modifed polymer for ecofriendly corrosion inhibition of copper in aqueous chloride environment, Res. Chem. Intermed. 43 (2017) 5811–5828. https://doi.org/10.1007/s11164-017-2964-x

[48]   K. Wan, P. Feng, B. Hou, Y. Li, Enhanced corrosion inhibition properties of carboxymethyl hydroxypropyl chitosan for mild steel in 1.0 M HCl solution, RSC Adv. 6 (2016) 77515–77524. https://doi.org/10.1039/C6RA12975G

[49]   J. Haque, V. Srivastava, D.S. Chauhan, H. Lgaz, M.A. Quraishi, Microwave-induced synthesis of chitosan schiffbases and their application as novel and green corrosion inhibitors: experimental and theoretical approach, ACS Omega. 3 (2018) 5654–5668. https://doi.org/10.1021/acsomega.8b00455

[50]   P.B. Raja, M.G. Sethuraman, Natural products as corrosion inhibitor for metals in corrosive media — A review, Mater. Lett. 62 (2008) 113–116. https://doi.org/10.1016/j.matlet.2007.04.079

[51]   K. Dahmani, M. Galai, M. Cherkaoui, A. El Hasnaoui, A. El Hessni, Cinnamon essential oil as a novel eco-friendly corrosion inhibitor of copper in 0.5 M Sulfuric Acid medium, J. Mater. Environ. Sci. 8 (2017) 1676–1689.

[52]   F. Mounir, S. El Issami, L. Bazzi, R. Salghi, N. Abidi, S. Jodeh, L. Bazzi, A. Eddine, Green approach to corrosion inhibition of copper by two oils of Argania Spinosa (L.) in phosphoric acid, J. Mater. Environ. Sci. 6 (2015) 2066–2075.

[53]   L. Palou, A. Ali, E. Fallik, G. Romanazzi, GRAS, plant-and animal-derived compounds as alternatives to conventional fungicides for the control of postharvest diseases of fresh horticultural produce, Postharvest Biol. Technol. 122 (2016) 41–52. https://doi.org/10.1016/j.postharvbio.2016.04.017

[54]   S. Marzorati, L. Verotta, S. P. Trasatti, Green corrosion inhibitors from natural sources and biomass wastes, Molecules. 48 (2019) 1-24. https://doi.org/10.3390/molecules24010048

[55]   M. Nasrollahzadeh, S.M. Sajadi, M. Khalaj, Green synthesis of copper nanoparticles using aqueous extract of the leaves of Euphorbia esula L and their catalytic activity for ligand-free Ullmann-coupling reaction and reduction of 4-nitrophenol, RSC Adv. 4 (2014) 47313–47318. https://doi.org/10.1039/C4RA08863H

[56]   J. Seo, S. Lee, M.L. Elam, S.A. Johnson, J. Kang, B.H. Arjmandi, Study tofind the best extraction solvent for use with guava leaves (Psidium guajava L.) for high antioxidant efficacy, Food Sci. Nutr. 2 (2014) 174–180. https://doi.org/10.1002/fsn3.91

[57]   O. Nkuzinna, M. Menkiti, O. Onukwuli, Inhibition of copper corrosion by acid extracts of Gnetum africana and Musa acuminate peel, Int. J. Multidiscip. Sci. Eng. 2 (2011) 2045–7057.

[58]   C. Verma, E. E. Ebenso, I. Bahadur, M. A. Quraishi, An overview on plant extracts as environmental sustainable and green corrosion inhibitors for metals and alloys in aggressive corrosive media, J. Mol. Liq. 266 (2018) 577–590. https://doi.org/10.1016/j.molliq.2018.06.110

[59]   B.E. A. Rani, B.B.J. Basu, Green Inhibitors for Corrosion Protection of Metals and Alloys: An Overview, Int. J. Corros. 2012 (2012) 1-15. https://doi.org/10.1155/2012/380217

[60]   M.P. Savita, N. Chaubey, S. Kumar, V.K. Singh, M.M. Singh, Strychnos nuxvomica, Piper longumand Mucuna pruriens, seed extracts as eco-friendly corrosion inhibitors for copper in nitric acid, RSC Adv. 6 (2016) 95644–95655. https://doi.org/10.1039/C6RA16481A

[61]   C. Rahal, M. Masmoudi, R. Abdelhedi, R. Sabot, M. Jeannin, M. Bouaziz, P. Refait, Olive leaf extract as natural corrosion inhibitor for pure copper in 0.5 M NaCl solution: a study by voltammetry around OCP, J. Electroanal. Chem. 769 (2016) 53–61. https://doi.org/10.1016/j.jelechem.2016.03.010

[62]   A.E.-A.S. Fouda, S.H. Etaiw, D.M.A. El-Azziz, O.A. Elbaz, Synergistic effect of barium chloride on corrosion inhibition of copper by aqueous extract of lupine seeds in nitric acid, Int. J. Electrochem. Sci. 12 (2017) 5934–5950. https://doi.org/10.20964/2017.07.08

[63]   C.A. Loto, R.T. Loto, A.P.I. Popoola, Inhibition effect of extracts of carica papaya and camellia sinensis leaves on the corrosion of duplex (α β) brass in 1m nitric acid, Int. J. Electrochem. Sci. 6 (2011) 4900-4914.

[64]   N. Raghavendra, J.I. Bhat, Application of green products for industrially important materials protection: An amusing anticorrosive behavior of tender arecanut husk (green color) extract at metal-test solution interface, Measurement. 135 (2019) 625–639. https://doi.org/10.1016/j.measurement.2018.12.021

[65]   F. Wedian, M.A. Al-Qudah, A.N. Abu-Baker, The effect of Capparis spinosa L. Extract as a green inhibitor on the corrosion rate of copper in a strong alkaline solution, Port. Electrochimica Acta. 34 (2016) 39-51. https://doi.org/10.4152/pea.201601039

[66]   R. Oukhrib, S. E. Issami, B. E. Ibrahimi, K. E. Mouaden, L. Bazzi, L. Bammou, A. Chaouay, R. Salghi, S. Jodeh, B. Hammouti, A. Amin-Alami, Ziziphus lotus as green inhibitor of copper corrosion in natural sea water, Port. Electrochim Acta. 35 (2017) 187–200. https://doi.org/10.4152/pea.201704187

[67]   A. Jmiai, B.E. Ibrahimi, A. Tara, M. Chadili, S.E. Issami, O. Jbara, A. Khallaayoun, L. Bazzi, Application ofZizyphus Lotuse- pulp of Jujube extract as green and promising corrosion inhibitor for copper in acidic medium, J. Mol. Liq. 268 (2018) 102–113. https://doi.org/10.1016/j.molliq.2018.06.091

[68]   F. Wedian, M.A. Al-Qudah, G.M. Al-Mazaideh, Corrosion inhibition of copper by Capparis spinosaL. extract in strong acidic medium: experimental and density functional theory, Int. J. Electrochem. Sci. 12 (2017) 4664–4676. https://doi.org/10.20964/2017.06.47

[69]   B. El Ibrahimi, K. El Mouaden, A. Jmiai, A. Baddouh, S. El Issami, L. Bazzi, M. Hilali, Understanding the influence of solution's pH on the corrosion of tin in saline solution containing functional amino acids using electrochemical techniques and molecular modeling, Surf. Interfaces 17 (2019) 100343. https://doi.org/10.1016/j.surfin.2019.100343

[70]   B. El Ibrahimi, A. Jmiai, K. El Mouaden, A. Baddouh, S. El Issami, L. Bazzi, M. Hilali, Effect of solution's pH and molecular structure of three linear α-amino acids on the corrosion of tin in salt solution: A combined experimental and theoretical approach, J. Mol. Struct. 1196 (2019) 105-118. https://doi.org/10.1016/j.molstruc.2019.06.072

[71]   B. El Ibrahimi, A. Jmiai, K. El Mouaden, R. Oukhrib, A. Soumoue, S. El Issami, L. Bazzi, Theoretical evaluation of somea-amino acids for corrosion inhibition of copper in acidic medium: DFT calculations, Monte Carlo simulations and QSPR studies, J. King Saud Univ. 32 (2020) 163-171. https://doi.org/10.1016/j.jksus.2018.04.004

[72]   M.P. Savita, N. Chaubey, V.K. Singh, M.M. Singh, Eco-friendly inhibitors for copper corrosion in nitric acid: theoretical and experimental evaluation, Metall. Mater. Trans. B 47 (2016) 47–57. https://doi.org/10.1007/s11663-015-0488-6

[73]   A.Y. El-Etre, M. Abdallah, Z.E. El-Tantawy, Corrosion inhibition of some metals using lawsonia extract, Corros. Sci. 47 (2005) 385–395. https://doi.org/10.1016/j.corsci.2004.06.006

[74]   R. Ouache, H. Harkat, P. Pale, K. Oulmi, Phytochemical compounds and anti-corrosion activity of Veronica rosea, Nat. Prod. Res. 33 (9) (2019) 1374-1378. https://doi.org/10.1080/14786419.2018.1474464

[75]   L. Vrsalovic, S. Gudic, D. Gracic, I. Smoljko, I. Ivanic, M. Kliskic, E.E. Oguzie, Corrosion protection of copper in sodium chloride solution using propolis, Int. J. Electrochem. Sci. 13 (2018) 2102–2117. https://doi.org/10.20964/2018.02.71

[76]   P.S. Pratihar, M.P.S. Verma, A. Sharma, Capparis decidua seeds: potential green inhibitor to combat acid corrosion of copper, Rasayan J. Chem. 8 (2017) 411–421.

[77]   Z. Chen, H. Cen, L. Wei, Y. Cao, X. Guo, Inhibitory action of garlic extract in the corrosion of copper under thin electrolyte layers, Surf. Rev. Lett. 1850128 (2017) 1-9. https://doi.org/10.1142/S0218625X18501287

[78]   A.R. Hajipour, F. Refiee, Recent Progress in Ionic Liquids and their Applications in Organic Synthesis, Org. Prep. Proced. Int. 47 (2015) 249–308. https://doi.org/10.1080/00304948.2015.1052317

[79]   Suresh, J.S. Sandhu, Recent advances in ionic liquids: green unconventional solvents of this century: part I, Green Chem. Lett. Rev. 4 (2011) 289–310. https://doi.org/10.1080/17518253.2011.572294

[80]   S. Zhang, N. Sun, X. He, X. Lu, X. Zhang, Physical properties of ionic liquids: database and evaluation, J. Phys. Chem. Ref. Data 35 (2006) 1475–1517. https://doi.org/10.1063/1.2204959

[81]   H. Zhao, Innovative applications of ionic liquids as"green"engineering liquids, Chem. Eng. Commun. 193 (2006) 1660–1677. https://doi.org/10.1080/00986440600586537

[82]   A. Fernicola, B. Scrosati, H. Ohno, Potentialities of ionic liquids as new electrolyte media in advanced electrochemical devices, Ionics 12 (2006) 95–102. https://doi.org/10.1007/s11581-006-0023-5

[83]   T. Tsuda, C.L. Hussey, Electrochemical applications of room-temperature ionic liquids, Electrochem. Soc. Interface 16 (2007) 42–49. https://doi.org/10.1149/2.F05071IF

[84]   C. Verma, E.E. Ebenso, M.A. Quraishi, Ionic liquids as green and sustainable corrosion inhibitors for metals and alloys: An overview, J. Mol. Liq. 233 (2017) 403–414. https://doi.org/10.1016/j.molliq.2017.02.111

[85]   N.V. Likhanova, M.A. Domínguez-Aguilar, O. Olivares-Xometl, N. Nava-Entzana, E. Arce, H. Dorantes, The effect of ionic liquids with imidazolium and pyridinium cations on the corrosion inhibition of mild steel in acidic environment, Corros. Sci. 52 (2010) 2088–2097. https://doi.org/10.1016/j.corsci.2010.02.030

[86]   X. Zheng, S. Zhang, W. Li, M. Gong, L. Yin, Experimental and theoretical studies of two imidazolium-based ionic liquids as inhibitors for mild steel in sulfuric acid solution, Corros. Sci. 95 (2015) 168–179. https://doi.org/10.1016/j.corsci.2015.03.012

[87] L. Feng, S. Zhang, Y. Qiang, Y. Xu, L. Guo, L. H. Madkour, S. Chen, Experimental and theoretical investigation of thiazolyl blue as a corrosion inhibitor for copper in neutral sodium chloride solution, Materials 11 (2018) 1042. https://doi.org/10.3390/ma11061042

[88] A. Bousskri, R. Salghi, A. Anejjar, M. Messali, S. Jodeh, O. Benali, M. Larouj, I. Warad, O. Hamed, B. Hammouti, The inhibition effect of 1-pentyl pyridazinium bromide towards copper corrosion in phosphoric acid containing chloride, Port. Electrochimica Acta 34 (2016) 1-21. https://doi.org/10.4152/pea.pea.201601001

[89] A. Bousskri, R. Salghi, A. Anejjar, S. Jodeh, M.A. Quraishi, M. Larouj, H. Lgaz, M. Messali, S. Samhan, M. Zougagh, Pyridazinium-based ionic liquids as corrosion inhibitors for copper in phosphoric acid containing chloride: electrochemical, surface and quantum chemical comparatives studies, Der. Pharma. Chemica. 8 (2016) 67-83.

[90] Y. Qiang, S. Zhang, L. Guoc, X. Zheng, B. Xiang, S. Chen, Experimental and theoretical studies of four allyl imidazolium-based ionic liquids as green inhibitors for copper corrosion in sulfuric acid, Corros. Sci. 119 (2017) 68–78. https://doi.org/10.1016/j.corsci.2017.02.021

[91] G. Vastag, A. Shaban, M. Vraneš, A. Tot, S. Belić, S. Gadžurić, Influence of the N-3 alkyl chain length on improving inhibition properties of imidazolium-based ionic liquids on copper corrosion, J. Mol. Liq. 264 (2018) 526–533. https://doi.org/10.1016/j.molliq.2018.05.086

[92] G.L.F. Mendonça, S.N. Costa, V.N. Freire, P.N.S. Casciano, A.N. Correia, P.d. Lima-Neto, Understanding the corrosion inhibition of carbon steel and copper in sulphuric acid medium by amino acids using electrochemical techniques allied to molecular modelling methods, Corros. Sci. 115 (2017) 41-55. https://doi.org/10.1016/j.corsci.2016.11.012

[93] H.H.A. Rahman, A.H.E. Moustafa, M.K. Awad, Potentiodynamic and quantum studies of some amino acids as corrosion inhibitors for copper, Int. J. Electrochem. Sci. 7 (2012) 1266–1287.

[94] G.M.A. El-Hafez, W.A. Badawy, The use of Cysteine, N-acetyl cysteine and Methionine as environmentally friendly corrosion inhibitors for Cu–10Al–5Ni alloy in neutral chloride solutions, Electrochim. Acta 108 (2013) 860–866. https://doi.org/10.1016/j.electacta.2013.06.079

[95] S. Varvara, I. Rotaru, M. Popa, R. Bostan, M. Glevitzky, L. Muresan, Environmentally-safe corrosion inhibitors for the protection of bronzes against

corrosion in acidic media, Chem. Bull. "POLITEHNICA" Univ. (Timisoara) 55 (2010) 156-161.

[96]   W. A. Badawy, K. M. Ismail, A. M. Fathi, Corrosion control of Cu–Ni alloys in neutral chloride solutions by amino acids, Electrochim. Acta 51 (2006) 4182–4189. https://doi.org/10.1016/j.electacta.2005.11.037

[97]   A.S. Fouda, Y.M. Abdallah, G.Y. Elawady, R.M. Ahmed, Electrochemical study on the efectively of Hyoscyamus muticusextract as a green inhibitor for corrosion of copper in 1 M HNO₃, J. Mater. Environ. Sci. 5 (2015) 1519–1531.

[98]   S.M. Ali, H.A.A. lehaibi, The inhibitive performance of fenugreek for corrosion of copper and nickel in sulfuric acid, Int. J. Electrochem. Sci. 11 (2016) 953–966.

[99]   Y.M. Abdallah, K. Shalabi, Comprehensive study of the behavior of copper inhibition in 1 M HNO₃ by Euphorbia helioscopialinn. extract as green inhibitor, Prot. Met. Phys. Chem. 51 (2015) 275–284. https://doi.org/10.1134/S2070205115020021

[100] M.A. Deyab, Egyptian licorice extract as a green corrosion inhibitor for copper in hydrochloric acid solution, J. Ind. Eng. Chem. 22 (2015) 384–389. https://doi.org/10.1016/j.jiec.2014.07.036

[101] H.S. Gadow, M.M. Motawea, H.M. Elabbasy, Investigation of myrrh extract as a new corrosion inhibitor for a-brass in 3.5% NaCl solution polluted by 16 ppm sulfde, RSC Adv. 7 (2017) 29883–29898. https://doi.org/10.1039/C7RA04271J

[102] P.D. Rani, S. Selvaraj, Emblica officinalis (AMLA) leaves extract as corrosion inhibitor for copper and its alloy (Cu-27Zn) in natural sea water, Arch. Appl. Sci. Res. 2 (2010) 140–150.

Sustainable Corrosion Inhibitors
Materials Research Foundations107 (2021) 204-221

Materials Research Forum LLC
https://doi.org/10.21741/9781644901496-9

Chapter 9

Case-Studies on Green Corrosion Inhibitors

M. Ramesh[1*], L. Rajeshkumar[2]

[1]Department of Mechanical Engineering, KIT-Kalaignarkarunanidhi Institute of Technology, Coimbatore-641402, Tamil Nadu, India

[2]Department of Mechanical Engineering, KPR Institute of Engineering and Technology, Coimbatore-641407, Tamil Nadu, India

*mramesh97@gmail.com

Abstract

Corrosion in metals and its alloys is an inevitable phenomenon but can be controlled by suitable classical methods like process control, cathode protection, surface treating methods, impurity reduction in metals and addition of metals to form alloys. Nevertheless, the employment of corrosion inhibitors is still a noteworthy and simplest of all the above processes in protecting the metals and alloys especially in acidic media. Protection of metals against corrosion not only prevents corrosion but also is beneficial in terms of money loss as far as industrial equipment, surfaces and vessels are concerned. Since the use of organic and inorganic inhibitors are highly discouraged due to their high cost and toxicity, necessity has adequately aroused the development of corrosion inhibitors which are natural and green. Trends, nowadays, focussed in controlling corrosion in various metals and alloys through green corrosion inhibitors consisting of natural elements alone. In contrast to the inorganic inhibitors, green corrosion inhibitors are characterized by biodegradability, low cost and meagre toxicity. Several researchers are now turning themselves towards the research of green inhibitors which are of no threat to humans and the ecosystem. The current discussion is focussed on the fundamentals of corrosion, corrosion inhibition, materials used for it and case studies of green inhibitors used for corrosion control in various conventional and monolithic metals.

Keywords

Corrosion, Green Inhibitors, Inhibition, Biodegradability, Plant Extracts

Contents

1. Introduction

When a reaction betweben metals and alloys with the environment takes place then it could be termed as corrosion. Usually at the time of corrosion the chemical or electrochemical reaction of the metals and alloys, used in different environments and applications, end up in the depletion of the top layer of the substrate due to its contact with acidic or aqueous environments. Corrosion of metal results in the loss of the metal considerably due to its environment. It is also a fact that corrosion is a never ending problem that requires sole attention but could not be eliminated completely. Its priority of addressing is very important such that it has direct influence over the conservation, safety and economic aspects in numerous applications like natural, medical, metallurgical and chemical engineering. Corrosion also affects the design of mechanical elements in various sizes and functions and also their lifespan. Corrosion induces heavy losses in terms of finance by directly affecting the safety, product loss, leakage, renewal and replacement of affected metals and alloy structures and ultimately resulting in environmental pollution and loss of efficiency. Depletion of metal substrate is often associated with heavy loss in productivity because of the faulty operation of corroded products and industrial instruments due to their presence in aqueous medium. Corrosion becomes one of the main problems that must be eliminated to avoid such harmful effects and if left unobserved, may result in equipment failure by all means [1].

National Association of Corrosion Engineers (NACE) performed an analysis in the year 1998 and concluded that a huge sum of 276 billion dollars was pumped into the economy for controlling corrosion which roughly approximates to 3.1 % of GDP of the USA. The amount magnified to 2.2 trillion dollars during the year 2011 for the US. During the Global Corrosion Summit that took place in New Delhi during the year 2011, it was declared that an amount of Rs. 2 lakh crore was spent on corrosion. Aforementioned data were possibly outdated, recent analysis of NACE during the year 2017 revealed that the global corrosion cost was 2.5 trillion US dollars approximately and it holds upto 3.4 % of global GDP [2]. A survey of NACE showed that approximate corrosion cost in South Africa was also approximately 130 billion US dollars approximately. All these corrosion costs could be easily reduced by 15 % to 35 % if the current prevention methods of corrosion are appropriately employed. Such prevention may also contribute for rise in GDP of all countries and thus globally.

Corrosion could easily and practically be prevented which is rather simpler than purging completely. Phenomenon of corrosion is slow until the top or protective passive layer of the metal gets worn off and then gets accelerated. Process of corrosion takes place along with numerous reactions between the metal substrate and its surrounding. For instance, the reaction could possibly be the local pH changes, oxide formation, electrochemical potential or/and metal cation penetration into coating of the substrate metal. Prevention of these reactions is the key function of an inhibitor. Specifically, while using acid solution in industries for functions like acid descaling, oil well acidizing, acid cleaning and acid pickling, the above reactions could be kept within limits by the use of a proper inhibitor and protect the metallic substrates [3]. Corrosion may not only end up in degradation or metal loss but also produces effects like resource wastage, reduction in yield, contamination of metal leading to loss of product, elevated cost of maintenance, safety hazards and slowing down the scientific progress of a nation. Larger structures like nuclear waste dispensation facilities, pipelines, tank trucks, storage tanks, ships, hazardous materials transport units and wagons would also be affected by corrosion heavily. Corrosion abates these structures and even compromises public safety and wellbeing.

Factors like deformation temperature of a metal substrate during its preparation, composition of the alloy, structure of metallic grain influences the affinity of a metal towards corrosion. Since the mechanism of corrosion is dependent on various factors like working metal, it's environment and its interaction with the surrounding, understanding of corrosion mechanism is considered to be mystical and hence prevention of corrosion according to the environment would be more feasible than trying to eliminate it. Some

Sustainable Corrosion Inhibitors   Materials Research Forum LLC
Materials Research Foundations 107 (2021) 204-221   https://doi.org/10.21741/9781644901496-9

factors which are to be addressed for corrosion prevention includes presence of electrolyte, sulphur dioxide, carbon dioxide, air and moisture [4]. Various types of corrosion, depending on the above said factors, are briefly described in Table 1.

Table 1.   Corrosion Classification [5].

Type	Narrative
Uniform corrosion	Corrosion takes place over the entire surface reducing its overall thickness
Galvanic corrosion	Takes place in between various metals having different corrosion potential in presence of an electrolyte
Pitting Corrosion	Induced on a surface at some specific points due to random attacks resulting in formation of pits. Develops as the undamaged surface acts as cathode while the pits acts as cathode
Stress corrosion cracking	Arises when a corrosive environment is acted upon stresses
Corrosion fatigue	Occurs due to combined action of corrosion and load reversal
Intergranular corrosion	Takes place mostly at the metal grain boundaries
Crevice corrosion	Occurs due to entrapment of corrosive liquids within the metal gaps due to concentration corrosion
Filiform corrosion	Occurs in surfaces of metals coated with organic film
Erosion corrosion	Occurs on surface of metals due to flow of corrosive liquids containing abrasives
Fretting corrosion	Another form of erosion corrosion which occurs due to metal fretting

2.   Corrosion inhibitors

An inhibitor is the combination of elements supplemented in smaller proportions to shield the exposed metallic surface from a corrosion atmosphere and it reduces or eliminates the corrosion in metals [6]. Corrosion inhibitor is said to be a chemical agent of preventing corrosion when it is present in the medium of corrosion at a concentration which does not pose much change in the other corrosive agent concentration. Rate of corrosion depends on the acidic-basic nature, agents of the environment, moisture and micro-dust particles.

Sustainable Corrosion Inhibitors   Materials Research Forum LLC
Materials Research Foundations**107** (2021) 204-221   https://doi.org/10.21741/9781644901496-9

Inhibitor substances reduce the rate of corrosion by absorbing the molecules and ions from the substrate metal [7]. By the formation of a passivating layer around the molecules, the interaction of the metal substrate with the surrounding environment is reduced. Usually corrosion inhibitors are added to metal substrates in ppm concentrations and these elements can have adsorption of substrate atoms at the physical and chemical levels which hinders the direct contact between corrosion agents and metal surfaces. Process of working of corrosion inhibitors is depicted in Fig. 1 [8].

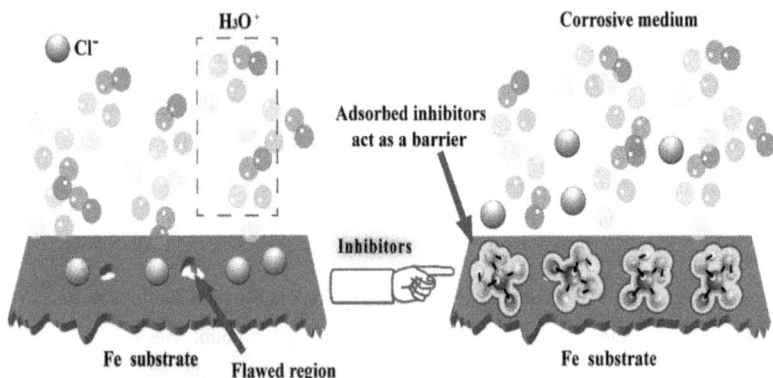

Figure 1 Corrosion inhibition mechanism [8].

Currently due to high toxic nature and being a foe to environments, uses of organic and inorganic inhibitors are reduced widely and the uses of green corrosion inhibitors are encouraged. These green inhibitors are the extracts from various parts of a plant and are environmentally friendly as they are not constituted with heavy metallic or toxic compounds. Green inhibitors usually possess biocompatibility with the environment and nature [9]. The following are the ways by which these inhibitors reduce the rate of corrosion: i) Ions/molecules of the inhibitor elements gets adsorbed by the substrate of the metal, ii) Reaction with cathode and/or anode may be increased or decreased, iii) slowing down the rate of diffusion of the reactant into the metal surface, iv) minimizing the electrical confrontation of the metal substrate, v) choosing an easily applicable green inhibitor which has an advantage of in situ processing. Besides these, choice of inhibitor plays a significant role which could be done by considering factors like availability, cost, environmental safety and flora type [10].

Such choice of inhibitors may orient the researcher to consider the following facts: i) Inhibitor material has to act effectively even at minimal concentrations and should impede dissolution of metal, ii) Inhibitor should have an indifference towards the intake of hydrogen along with thermal stability and inertness towards chemicals, iii) Attributes to acts as surfactant and foaming properties should be better, iv) Inhibitor chosen should be easily available, cheap and non-toxic. When the above factors were considered simultaneously at the time of choosing, usage of green inhibitors becomes more feasible and reliable [11].

3.   Use of green corrosion inhibitors

Corrosion of metals can be readily prevented by the use of green corrosion inhibitors. Currently, the inhibitors in existence were prepared from carbon compounds with long chain and aromatic heteroatoms or from raw materials of low quality. Such inhibitors were highly toxic in nature which slowly promoted the use of green corrosion inhibitors. Inhibitors like Delonix regia and Opuntia extracts prevented corrosion of aluminium in HCL medium [12] while corrosion of aluminium and magnesium alloy in 3 % NaCl medium was prevented by extracts prepared from rosemary leaves. Extracts of natural honey inhibited the copper corrosion while Khillah seed extracts had an appreciable inhibitive effect over 316 stainless steel in acidic environments [13]. Extracts from African bush pepper rendered ethanol which inhibited the corrosion of mild steel and the neem leaves extract prevented corrosion of the same metal in sulphuric acid environments. Corrosion inhibition of steel in acidic environments was appreciable given by extracts like Auforpio turkiale sap, Papaia, Azadirachta indica, Poinciana pulcherrima, Calotropis procera, Cassia occidentalis, Datura stramonium seeds and Papaya with an overall efficiency of 95 %. Corrosion inhibition mechanism in all the above cases was due to the formation of highly insoluble complexes with iron as cations. Better corrosion inhibition of the above plants was purely due to the protein hydrolysis reaction [14-20]. Measurements of corrosion inhibition were normally made by potentiodynamic polarization studies and weight loss methods. Fig. 2 depicts the possibilities of using plants for the extraction of green inhibitors.

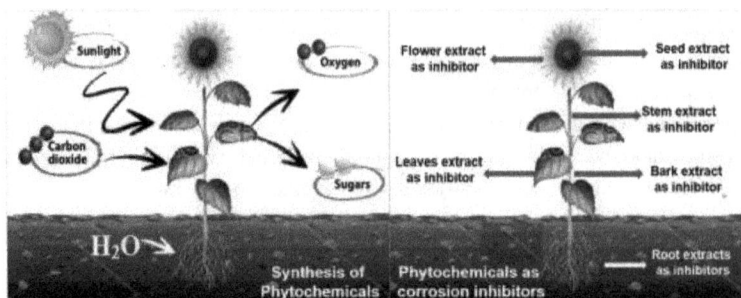

Figure 2 Green corrosion inhibitors from plants [21].

4.   Plant extracts as corrosion inhibitors

4.1   Corrosion inhibitors for low carbon steel

Nik et al. [22] performed experiments on Henna extracts and exhibited that the Henna extracts were behaving mostly as cathodic side inhibitor and as a mixed type. Most of the researchers found the extracts of Henna to be used as inhibitors for aluminium alloy 5083. The inhibitor effect of henna is mainly due to one of its constituents namely Lawsone. Singh et al. [23] considered Bacopa monnieri as a corrosion inhibitor for aluminium metal in an alkaline environment and measured the rate of corrosion by weight reduction estimations and polarization techniques. Experiments revealed that the inhibiting effect of Bacopa monnieri extract has high efficiency and the comparison between the rate of corrosion between weight loss and polarization techniques rendered almost the same results. Bacopa monnieri extract behaved as a mixed-type inhibitor on aluminium metal in alkaline environments and the reaction between cathode and anode was avoided by using it. These facts were observed from potentiodynamic polarization test results. Johnsirani et al. [24] examined the inhibition efficiency (IE) of aqueous extracted Henna leaves by applying them upon carbon steel placed in sea water. Experiments were conducted by weight loss method and the results depicted that the compound comprising 25 ppm of Zn^{2+} mixed with 8 ml of henna extract possesses an efficiency of 94%. Same experiments were repeated by Polarization method and the results were almost aligned revealing the inhibitor to be a mixed type. Characterization techniques like SEM, FTIR, EIS and AFM were used to analyse the metal surface and it was found that a protective film was formed by the inhibitor over the metal substrate.

Nour et al. [25], analysed the corrosion inhibition of carbon steel by employing various concentrations of aqueous extracted henna leaves in an acidic solution of HCl. Corrosion rate was determined by weight loss and potentiodynamic polarization techniques. Besides this the corrosion behaviour of carbon steel at a temperature range of 20 to 60°C was also analysed. IE increased with inhibitor concentration while it decreased with raise in temperature. The physical adsorption mechanism of carbon steel with henna extracts was braced by the activation of free energies. The adsorption on the carbon steel surface was compatible with the Langmuir's adsorption isotherm, endothermic and spontaneous. Potentiodynamic polarization measurements demonstrate that the henna extraction shows as mixed inhibitor. Surface analysis and protective film characterization was carried out by using; Fourier transforms infrared (FTIR), spectroscopy, scanning electron microscopy (SEM), energy dispersive X-ray (EDX), and X-ray diffraction (XRD) analysis. It could be observed from the experiments that the order of occurrence was Langmuir adsorption isotherm and then adsorption.

4.2   Corrosion inhibitors for mild steel

A seaweed termed as Leucaena leucocephala (commonly known as River tamarind) was reported to have better inhibition towards corrosion for the components in seawater. The statistics taken by the transportation department of USA, within a span of six years from 1994 to 1999, reported that almost 26 % of mild steel pipeline accidents occurred mainly because of corrosion [26]. Dry powder of river tamarind was bought from Secret Barn Sdn. Bhd. and used directly. It was soaked in ethanol and filtered using a vacuum pump. The powder was then evaporated using rotovap and then stored in the refrigerator until used. Mild steel specimens were cut into 25 mm × 25 mm × 3 mm and polished using sandpaper until mirror finish. They were finally cleaned using acetone and stored in a desiccator for further use. Immersion test was carried out in seawater for 240 hours and the characteristics were studied using electron impedance spectroscopy (EIS) and SEM [27].

The coating produced by integrating the extract of river tamarind with a local paint proved to be successful and able to protect the mild steel against the damage of corrosion to a certain extent when immersed in seawater. Analysis via FTIR confirms the presence and the involvement of several functional groups such as hydroxyl and carbonyl groups which proposedly supplies the neutral molecules for the inhibition mechanism. Fig. 3 shows a plot of corrosion of mild steel specimens when immersed in sea water. The analysis conducted through electrochemical studies as shown in Fig. 4, indicate the efficiency of the coating increases with the increase of LLE concentration until a certain limitation or optimum value is achieved [28].

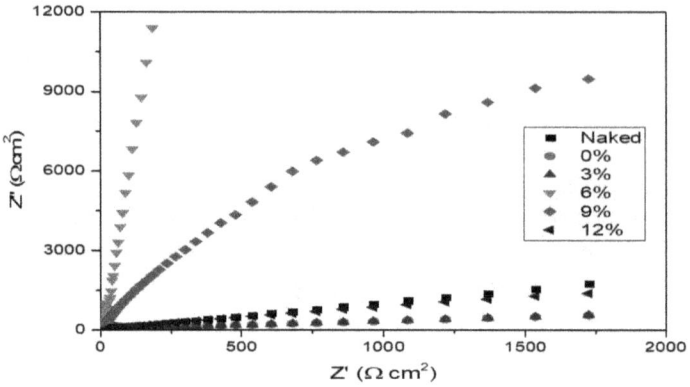

Figure 3 *Nyquist plot for coated and uncoated mild steels immersed in seawater [27].*

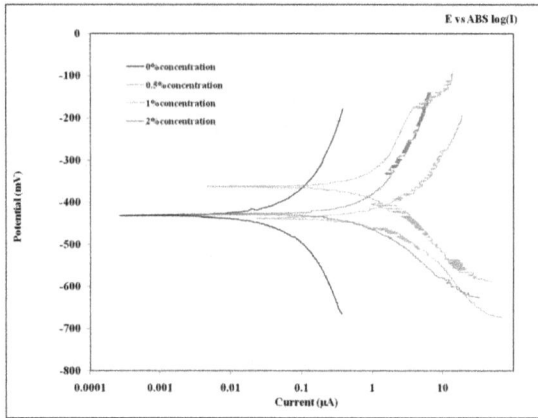

Figure 4 *Potentiodynamic polarization curves of low carbon steel in Henna solution [28].*

On the other hand, the study also implies that the coating acts as a mixed type coating where it reduces the progress of corrosion on both anodic and cathodic sites. Based on the observation via SEM, as shown in Fig. 5, the surface of the coating with the optimum concentration of LLE shows a smooth and well distributed layer with no sign of cracks and pores indicating the improvement in its capabilities when compare to the observation of the same coating without the presence of extract. This demonstrates a remarkable use of natural resources as a viable method in improving the present corrosion inhibitor [29].

Figure 5 SEM images of cross-sectional area of (a) freshly polished mild steel coupon, (b) naked mild steel and (c) mild steel after 240 hours of immersion in seawater [29].

4.3   Corrosion inhibitors for aluminium in corrosion media

Owing to a better technological impact arising from an uninterrupted necessity in industries, aluminium and its alloys have incessantly engrossed industrialists, metallurgists and researchers. They are applied in food packing, electrical transmission, thermal reservoirs and light-weight transport sectors. Aluminium and its alloys are characterized by a unique feature of formation of passivating oxide layer over their surface which makes them highly resistant towards oxidation. This layer is usually spontaneously formed either as anodic or cathodic film. But compounds containing alkali elements and concentrated base or acidic solutions easily penetrate within the passivating layer inducing corrosion. However, Neutral solutions pose no threat to the protective oxide layer over aluminium metal within a pH range of 4.0–8.5. Hence plant extracts or parts of any plants are believed to act as a corrosion inhibitor with high efficiency because they contain nearly analogous molecular and electronic structures of emblematic organic inhibitors and active pyto-chemcial compounds. These plant extracts are sturdily close at dynamic locations of aluminium at the time of corrosion and indorses anti-corrosion then. Borassus flabellifer (BF) is an eco-friendly and biodegradable fiber and the BF extracts comprises abundant hemicelluloses, cellulose and lignin contents [30]. These fiber extracts possess elemental functional groups that will strappingly stick to metal substrate and protect them from its corrosion in any acidic or in alkaline corrosive environment. Several parts of BF namely coir, fruits, shell, seeds, flowers, leaves and roots were derived from the plant and analysed by various researchers.

Analysis was also carried out on BF dust (BFD) derived out of water and methyl alcohol and was considered to be a biodegradable, minimal cost and environmental friendly inhibitor of corrosion for aluminium metals in hostile acid solutions. Furthermore, it was revealed from the experiments that the efficiency of corrosion inhibition surges with increase of BFD content and maximum IE was noted to be 67 % and 52 % in the case of BFD mined from methyl alcohol (BFDME) and water (BFDWE), respectively. Both the materials were derived at a similar concentration of 0.41 g/l. Likewise, polarization graphs portray that the extracts implement a mixed nature with somewhat dominant on adjacent cathode. Meanwhile, adsorption isotherm and thermodynamic calculations illustrates that materials were adsorbed by both physisorption and chemisorption on metallic aluminium surface which also obeys Langmuir adsorption isotherm. Micrograph and morphological studies by SEM portrays the ability of BFDME and BFDWE on corrosion inhibition on aluminium while kept in 1M H_2SO_4 and could be readily recommended for industrial use [31].

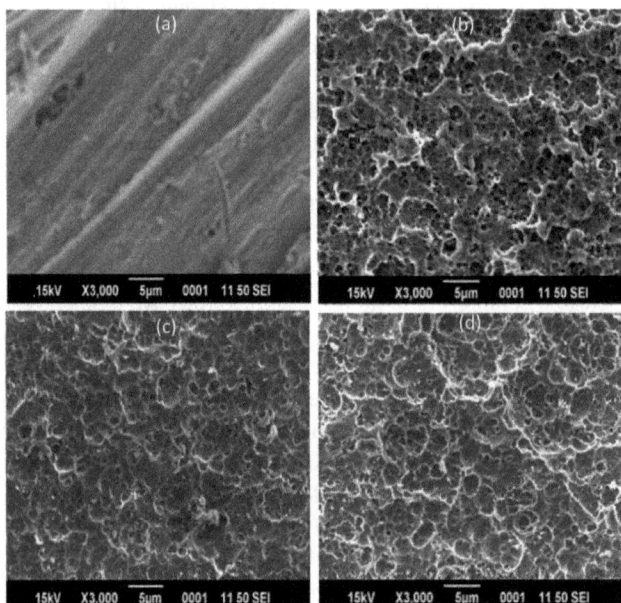

Figure 6 SEM image of aluminium: (a) before absorption; (b) after absorption in 1M H_2SO_4; (c) after absorption in 0.4 g/l BFDME+ 1M H_2SO_4; (d) after absorption in 0.4 g/l BFDWE+ 1M H_2SO_4 [31].

IE of BFDME and BFDWE was observed to increase with the increase in concentration of inhibitor added to the metal substrate. According to the weight-loss method test results, maximum IE was obtained as 66% and 51% for both of the above extracted materials at a temperature of 30°C in the same concentrations. SEM images also shows that inhibition of corrosion in aluminium alloys by the addition of BFDME and BFDWE was due to the formation of a passivating top layer on the metal substrate. The SEM micrographs of bare and inhibited aluminium specimens are shown in Fig. 6 [31].

4.4 Natural gums as corrosion inhibitors for metals

Gums are the natural elements obtained from the trunks or stem of various plants and they mostly contain polysaccharides which are soluble in water. When they are kept in water, due to their hydrophilic nature, they absorb water and get swollen up to form a gel like element. However they are insoluble in solvents and oils like alcohol, hydrocarbons and ether. The constituents of gum are normally complex and when they undergo hydrolysis reaction they render simple sugar compounds like glucuronic acid, mannose, arabinose and galactose. If the trunk or stem of a tree is injured pathologically in a gradual manner or by boring of insects or accidentally then the phenomenon is described as exudation. Gums are formed in the stems or root of trees at the time of exudation. When the seed endosperm portions were separated from them, then seed gums were formed. Apart from the application of gums towards corrosion inhibition they are also used in various other applications like pharmaceutical, food and other technical applications. Because of dominant stabilizing, emulsifying, suspending and thickening characteristics, gums are widely used in food and drinks applications in the food industry. On the other hand, gums are used as emulsifying agents for lotions and creams, tablet binding agents and in some other medical fields including dental in pharmaceutical industries. Above all, it could be seen that gums are excellent and effective corrosion inhibitors.

Eddy et al. [32] stated that gums were excellent corrosion inhibitors due to some of the reasons including: i) Gums form complex metal ions on the surface of the metals with the aid of their functional groups, ii) The so formed metal complexes protect the metal substrate from the causal agents of corrosion by inhabiting a large surface area thereby sheltering the surface, iii) Adsorption behaviour of gums are primarily due to presence of compounds like glucoprotein, sucrose, polysaccharides, arbinogalactan and oligosaccharides as these are constituted primarily with nitrogen and oxygen, iv) Adsorption mechanism of the gums were enhanced by the transfer of charge and electrons as they are rich in CO-OH organic groups, v) Usually like many other green inhibitors gums are also environmental friendly and low in toxicity. Some of the commonly used gums are Gum Arabic, Locust Bean gum, Guar gum, Albizia gum, Excudated gum Raphia hookeri, Dacroydes edulis, Ficus glumosa gum, Commiphora kestingii gum, exudated gum Pachylobus edulis, Ficus Benjamina gum, Anogessius leocarpus gum, Commiphora pendunculata gum, Ficus platyphylla gum, Ficus tricopoda gum, Daniella olieverri gum, Ficus thonningii gum, Khaya ivorensis gum, Gloriosa superba gum.

4.5   Ionic liquids as green corrosion inhibitors

Though iron and its alloys are cheap in cost and easy to produce, they are more prone to corrosion in acidic media especially during acid cleaning and the acid pickling process. Restrictions, due to ecological balance maintenance and stringent consciousness towards environment, are unavoidable when the toxic inhibitors of corrosion were used for metals. Hence depletion of ionic liquids to act as inhibitors for corrosion is emerging as an inevitable substitutive technique for the prevention of corrosion by green methods. These liquids have numerous better properties like high polarity, chemical stability, lower vapour pressure, lower melting point (lower that 100°C), high thermal stability and low toxicity render them to be an ideal alternative for conventional toxic inhibitors of corrosion. Fig. 7 depicts various applications of the ionic liquids for using them as green corrosion inhibitors [33]. The ionic liquids considered here inhibits corrosion in metallic substrates by adsorbing the molecules of metal obeying the Langmuir adsorption isotherm. Phenomenon of adsorption was supported by scanning electron microscopy, energy dispersive X-ray spectroscopy and atomic force microscopic analysis. Polarization analysis depicted that the ionic liquids considered had acted as an inhibitor of mixed type.

Copper and its alloys were broadly utilized for various applications such as coinages, electronics, ornamental, electricity and building construction industries and also in manufacturing of equipment owing to better corrosion resistance, electrical and thermal and mechanical properties [34]. Nonetheless, in occurrence of belligerent anionic elements like nitrates, sulphates and chlorides, copper and alloys may be severely attacked making them corrosion prone and ending up with loss of materials. Similarly, zinc and its alloys are widely used in chemical and petroleum industries [35]. But these industries are considered to have a constant fluctuation of pH values between maximum and minimum which results in degradation of mechanical properties due to corrosion. 1-butyl-3methylimidazoliumbis (trifluoromethyl-sulfonyl) imide, 1-hexyl-3methylimidazoliumbis (trifluoromethyl-sulfonyl)   imide,   and   1-propyl-2,3   methylimidazoliumbis (trifluoromethyl-sulfonyl) imide were the most preferably used synthesized green inhibitors for copper and zinc alloys. Results showed that both the ionic liquids acted as good corrosion inhibitors and their inhibition efficiency increases with increasing their concentrations.

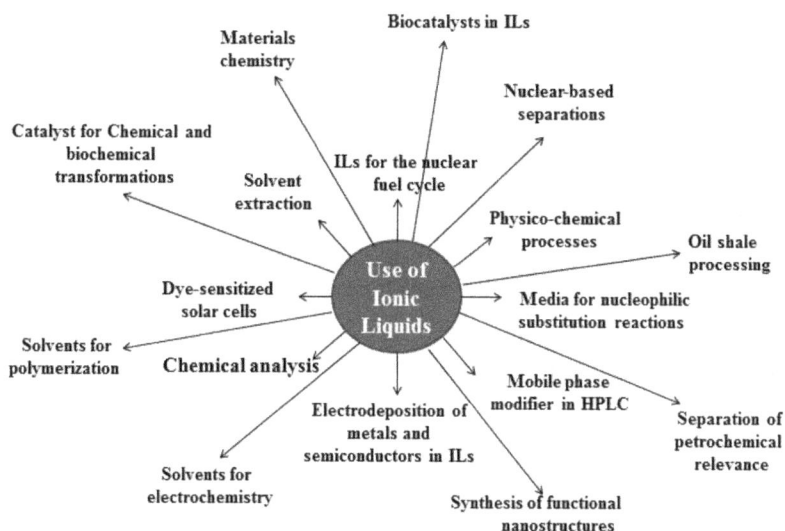

Figure 7. Applications of ionic liquids [33].

Conclusion

Efforts to prevent corrosion of metals are in force these days by employing green corrosion inhibitors. It could be stated from the above case studies that usage of plant excerpts as green inhibitors towards corrosion is much feasible and effective for a wide range of metals. When the metals were present in hydrochloric of sulphuric acid environments, then potentiodynamic polarization, weight loss and electrochemical impedance methods could be readily used to analyse the inhibition of corrosion. Though most of the inhibitors which are processed in the laboratories has exhibited better efficiency, their usage in industries is very limited due to their toxic nature. This shifted the focus of various researchers towards the processing and application of inhibitors that are environmental friendly. In spite of its successful processing, application of green inhibitors in real industrial sectors has to be explored more. Corrosive atmospheres involving harmful gases like hydrogen sulphide, carbon dioxide and sulphur dioxide are to be analysed more for the application of green inhibitors in such environments. All green inhibitors derived from plant oils are characterized by non-toxicity, renewable, cheap and high efficiency. Nevertheless efforts for finding out the compound responsible for the inhibition property in any plant are rarely reported. When the inhibitor is considered to be a mixture of compounds having better inhibition efficiency, they experience synergistic effects. In this way, such green inhibitors are exhibiting an efficiency of nearly 98%. Hence the excerpts of plants are good corrosion inhibitors that could be positively used on an industrial level. Yet concentration could be given further on the numerical modelling of the inhibition mechanism by green inhibitors which may be helpful in instituting the mechanism of inhibition in detail. It may also open many avenues for finding suitable alternatives.

References

[1]   M. Sangeetha, S. Rajendran, T.S. Muthumegala, A. Krishnaveni. Green corrosion inhibitors-an overview, Zastita Materijala. 52(1) (2011) 3-19.

[2]   J.T. Stephen, A. Adebayo, Inhibition of corrosion of mild steel in hydrochloric acid solution using akee apple seed extract. J. Fail. Anal. Prev. 18 (2018) 350–355. https://doi.org/10.1007/s11668-018-0431-7

[3]   H. Habeeb, H.M. Jwad, T.A. Luaibi, R.M. Abdullah, A.A. Dakhil, H. Kadhum, A.A. Al-Amiery, Case study on thermal impact of novel corrosion inhibitor on mild steel, Case Stud. Therm. Eng. 12 (2018) 64-68. https://doi.org/10.1016/j.csite.2018.03.005

[4]   B.E. Rani, B.J.B. Bharathi, Green inhibitors for corrosion protection of metals and alloys: an overview. Int. J. Corros. 2012 (2012) 380217. https://doi.org/10.1155/2012/380217

[5]   D.K. Singh, S. Kumar, G. Udayabhanu, R.P. John. 4(N,Ndimethylamino) benzaldehyde nicotinic hydrazone as corrosion inhibitor for mild steel in 1M HCl solution: An experimental and theoretical study, J. Mol. Liq. 216 (2016) 738–746. https://doi.org/10.1016/j.molliq.2016.02.012

[6]   K.K. Alaneme, S.J. Olusegun, O.T. Adelowo. Corrosion inhibition and adsorption mechanism studies of Hunteria umbellata seed husk extracts on mild steel immersed in acidic solutions, Alex. Eng. J. 55(1) 2016) 673–681. https://doi.org/10.1016/j.aej.2015.10.009

[7]   A. Harmaoui, H. Bourazmi, M.E. Fal, M. Boudalia, M. Tabyaoui, A. Guenbour, A. Bellaouchou, Y. Ramli, E.M. Essassi, Study of triazolotiazepinon as a corrosion inhibitor in 1M hydrochloric acid, Environ. Sci., 6(9) (2015) 2509-2519.

[8]   W. Belmaghraoui, A. Mazkour, H. Harhar, M. Harir, S. El-Hajjaji, Investigation of corrosion inhibition of C38 steel in 5.5 M H_3PO_4 solution using Ziziphus lotus oil extract: an application model, Anti-Corros. Meth. Mater. 66(1) (2019) 121-126. https://doi.org/10.1108/ACMM-02-2018-1901

[9]   R. Rajalakshmi, A. Prithiba, S. Leelavathi, An overview of emerging scenario in the frontiers of eco-friendly corrosion inhibitors of plant origin for mild steel, J. Chem. Acta 1(1) (2012) 6-13.

[10]   G. Ting, P. Su, X. Liu, J. Zou, X. Zhang, Y. Hu, A composite inhibitor used in oilfield: MA-AMPS and imidazoline. J. Petrol. Sci. Eng.102 (2013): 41-46. https://doi.org/10.1016/j.petrol.2013.01.014

[11]   F.J. Schwinn, Ergosterol biosynthesis inhibitors. An overview of their history and contribution to medicine and agriculture, Pestic. Sci. 15(1) (1984) 40-47. https://doi.org/10.1002/ps.2780150107

[12]   S. Abdoul-Azize, Potential benefits of jujube (Zizyphus Lotus L.) bioactive compounds for nutrition and health, J. Nutr. Metabol. (2016) 2867470. https://doi.org/10.1155/2016/2867470

[13]   H. Saufi, A. Al-Maofari, A. El Yadini, L. Eddaif, H. Harhar, S. Gharby, S. El-Hajjaji, Evaluation of vegetable oil of nigel as corrosion inhibitor for iron in NaCl 3% medium, J. Mater. Environ. Sci. 6(7) (2015) 1845-1849.

[14]   R.T. Vashi, N.I. Prajapati, Corrosion Inhibition of Aluminium in hydrochloric acid using bacopa monnieri leaves extract as green inhibitor, Int. J. ChemTech Res. CODEN (USA) IJCRGG 10(15) (2017) 221-231.

[15]   E.E. Oguzie, Corrosion inhibitive effect and adsorption behaviour of Hibiscus Sabdariffa extract on mild steel in acidic media, Port. Electrochem. Acta 26 (2008) 303–314. https://doi.org/10.4152/pea.200803303

[16]   E.E. Oguzie, K.L. Iyeh, A.I. Onuchukw, Inhibition of mild steel corrosion in acidic media by aqueous extracts from Garcinia kola seed. Bull. Electrochem. 22(2) (2006) 63–68.

[17]   M.S. Kumar, K. Sudesh, R. Ratnani, S.P. Mathur, Corrosion inhibition of aluminium by extracts of Prosopis cineraria in acidic media, Bull. Electrochem. 22(2) (2006) 69–74.

[18]   A.Y. El-Etre, M. Abdallah, Z.E. El-Tantawy, Corrosion inhibition of some metals using lawsonia extract, Corros. Sci. 47(2) (2005) 385–395. https://doi.org/10.1016/j.corsci.2004.06.006

[19]   A. Bouyanzer, L. Majidi, B. Hammout, Effect of eucalyptus oil on the corrosion of steel in 1 M HCl, Bull. Electrochem, 22(7) (2006) 321–324.

[20] M.J. Sanghvi, S.K. Shuklan, A.N. Misra, M.R. Padh, G.N. Mehta, Inhibition of hydrochloric acid corrosion of mild steel by aid extracts of Embilica officianalis,

Terminalia bellirica and Terminalia chebula, Bull. Electrochem. 13(8–9) (1997) 358–361.

[21]   P.B. Raja, M. Ismail, S. Ghoreishiamiri, J. Mirza, M.C. Ismail, S. Kakooei, A.A. Rahim, Reviews on corrosion inhibitors: a short view, Chem. Eng. Comm. 203(9) (2016) 1145-1156. https://doi.org/10.1080/00986445.2016.1172485

[22]   W.B.W. Nik, F. Zulkifli, R. Rosliza, M.M. Rahman, Lawsonia Inermis as green inhibitor for corrosion protection of aluminium alloy, Int. J. Mod. Eng. Res. 1(2) (2011) 723-728.

[23]   E.E. Singh, Ebenso, M.A. Quraishi, Stem extract of Brahmi (Bacopa monnieri) as green corrosion inhibitor for Aluminum in NaOH solution, Int. J. Electrochem. Sci. 7 (2012) 3409–3419.

[24] V. Johnsirani, J. Sathiyabama, S. Rajendran, A.S. Prabha, Inhibitory mechanism of carbon steel corrosion in sea water by an aqueous extract of henna leaves, Int. Sch. Res. Network ISRN Corros. (2012) 574321. https://doi.org/10.5402/2012/574321

[25]   A.H. Nour, S. El-Gendy, Thermodynamic, adsorption and electrochemical studies for corrosion inhibition of carbon steel by henna extract in acid medium, Egypt. J. Petrol. 22(1) (2013) 17-25. https://doi.org/10.1016/j.ejpe.2012.06.002

[26]   I. Adejoro, F. Ojo, S. Obafemi, Corrosion inhibition potentials of ampicillin for mild steel in hydrochloric acid solution, J. Taibah Univ. Sci., 9(2) (2015) 196–202. https://doi.org/10.1016/j.jtusci.2014.10.002

[27]   W.M.K.W.M. Ikhmal, M.F.M. Maria, W.A.W. Rafizah, W.N.W.M. Norsani, M.G.M. Sabri, Corrosion inhibition of mild steel in seawater through green approach using Leucaena leucocephala leaves extract, Int. J. Corros. Scale Inhib. 8(3) (2019) 628-643. https://doi.org/10.17675/2305-6894-2019-8-3-12

[28]   Al Hasan, N.H. Jasim, H.J. Alaradi, Z.A. Khadhim Al Mansor, A.H.J. Al-Shadood, The dual effect of stem extract of Brahmi (Bacopamonnieri) and Henna as a green corrosion inhibitor for low carbon steel in 0.5 M NaOH solution, Case Stud. Construct. Mater. 11 (2019) 00300. https://doi.org/10.1016/j.cscm.2019.e00300

[29]   D. Jeroundi, H. Elmsellem, S. Chakroune, R. Idouhli, A. Elyoussfi, A.E. Dafali, E.M. Hadrami, A. Ben-Tama, R.Y. Kandri. Physicochemical study and corrosion inhibition potential of dithiolo[4,5-b][1,4] dithiepine for mild steel in acidic medium, Environ. Sci. 7(11) (2016) 4024-4035.

[30]   S. Issaadi, T. Douadi, A. Zouaoui, S. Chafaa, M.A. Khan, G. Bouet, Novel thiophene symmetrical Schiff base compounds as corrosion inhibitor for mild steel in

acidic media, Corros. Sci. 53 (2011) 1484–1488.
https://doi.org/10.1016/j.corsci.2011.01.022

[31]   R.S. Nathiya, S. Perumal, V. Murugesan, V. Raj, Evaluation of extracts of Borassus flabellifer dust as green inhibitors for aluminium corrosion in acidic media, Mater. Sci. Semicond. Process. 104 (2019) 104674. https://doi.org/10.1016/j.mssp.2019.104674

[32]   N.O. Eddy, E.E. Ebenso, Adsorption and inhibitive properties of ethanol extracts of Musa sapientum peels as a green corrosion inhibitor for mild steel in H_2SO_4, Afr. J. Pure. Appl. Chem. 2(6) (2008) 46–54.

[33]   C. Verma, E.E. Ebenso, M.A. Quraishi. Ionic liquids as green and sustainable corrosion inhibitors for metals and alloys: An overview, 233 (2017) 403-414. https://doi.org/10.1016/j.molliq.2017.02.111

[34]   H. Tavakoli, T. Shahrabi, M.G. Hosseini. Synergistic effect on corrosion inhibition of copper by sodium dodecylbenzenesulphonate (SDBS) and 2-mercaptobenzoxazole, Mater. Chem. Phys. 109 (2008) 281-288. https://doi.org/10.1016/j.matchemphys.2007.11.018

[35]   M.A. Elmorsi, A.M. Hassanein, Corrosion inhibition of copper by heterocyclic compounds, J. Corros. Sci., 41 (1999) 2337-2352. https://doi.org/10.1016/S0010-938X(99)00061-X

Keyword Index

About the Editors

Dr. Inamuddin is working as Assistant Professor at the Department of Applied Chemistry, Aligarh Muslim University, Aligarh, India. He obtained Master of Science degree in Organic Chemistry from Chaudhary Charan Singh (CCS) University, Meerut, India, in 2002. He received his Master of Philosophy and Doctor of Philosophy degrees in Applied Chemistry from Aligarh Muslim University (AMU), India, in 2004 and 2007, respectively. He has extensive research experience in multidisciplinary fields of Analytical Chemistry, Materials Chemistry, and Electrochemistry and, more specifically, Renewable Energy and Environment. He has worked on different research projects as project fellow and senior research fellow funded by University Grants Commission (UGC), Government of India, and Council of Scientific and Industrial Research (CSIR), Government of India. He has received Fast Track Young Scientist Award from the Department of Science and Technology, India, to work in the area of bending actuators and artificial muscles. He has completed four major research projects sanctioned by University Grant Commission, Department of Science and Technology, Council of Scientific and Industrial Research, and Council of Science and Technology, India. He has published 186 research articles in international journals of repute and nineteen book chapters in knowledge-based book editions published by renowned international publishers. He has published 144 edited books with Springer (U.K.), Elsevier, Nova Science Publishers, Inc. (U.S.A.), CRC Press Taylor & Francis Asia Pacific, Trans Tech Publications Ltd. (Switzerland), IntechOpen Limited (U.K.), Wiley-Scrivener, (U.S.A.) and Materials Research Forum LLC (U.S.A). He is a member of various journals' editorial boards. He is also serving as Associate Editor for journals (Environmental Chemistry Letter, Applied Water Science and Euro-Mediterranean Journal for Environmental Integration, Springer-Nature), Frontiers Section Editor (Current Analytical Chemistry, Bentham Science Publishers), Editorial Board Member (Scientific Reports-Nature), Editor (Eurasian Journal of Analytical Chemistry), and Review Editor (Frontiers in Chemistry, Frontiers, U.K.) He is also guest-editing various special thematic special issues to the journals of Elsevier, Bentham Science Publishers, and John Wiley & Sons, Inc. He has attended as well as chaired sessions in various international and national conferences. He has worked as a Postdoctoral Fellow, leading a research team at the Creative Research Initiative Center for Bio-Artificial Muscle, Hanyang University, South Korea, in the field of renewable energy, especially biofuel cells. He has also worked as a Postdoctoral Fellow at the Center of Research Excellence in Renewable Energy, King Fahd University of Petroleum and Minerals, Saudi Arabia, in the field of polymer electrolyte membrane fuel cells and computational fluid dynamics of polymer electrolyte membrane fuel cells. He is a life member of the Journal of the Indian

Chemical Society. His research interest includes ion exchange materials, a sensor for heavy metal ions, biofuel cells, supercapacitors and bending actuators.

Dr. Mohd Imran Ahamed received his Ph.D degree on the topic "Synthesis and characterization of inorganic-organic composite heavy metals selective cation-exchangers and their analytical applications", from Aligarh Muslim University, Aligarh, India in 2019. He has published several research and review articles in the journals of international recognition. Springer (U.K.), Elsevier, CRC Press Taylor & Francis Asia Pacific and Materials Research Forum LLC (U.S.A). He has completed his B.Sc. (Hons) Chemistry from Aligarh Muslim University, Aligarh, India, and M.Sc. (Organic Chemistry) from Dr. Bhimrao Ambedkar University, Agra, India. He has co-edited more than 20 books with Springer (U.K.), Elsevier, CRC Press Taylor & Francis Asia Pacific, Materials Research Forum LLC (U.S.A) and Wiley-Scrivener, (U.S.A.). His research work includes ion-exchange chromatography, wastewater treatment, and analysis, bending actuator and electrospinning.

Dr. Mohammad Luqman has 12+ years of post-PhD experience in Teaching, Research, and Administration. Currently, he is serving as an Assistant Professor of Chemical Engineering in Taibah University, Saudi Arabia. Before joining here, he served as an Assistant Professor in College of Applied Science at A'Sharqiyah University, Oman, and in College of Engineering at King Saud University, Saudi Arabia. He served as a Research Engineer in SAMSUNG Cheil Industries, South Korea. Moreover, he served as a post-doctoral fellow at Artificial Muscle Research Center, Konkuk University, South Korea, in the field of Ionic Polymer Metal Composites for the development of Artificial Muscles, Robotic Actuators and Dynamic Sensors. He earned his PhD degree in the field of Ionomers (Ion-containing Polymers), from Chosun University, South Korea. He successfully served as an Editor to three books, published by world renowned publishers. He published numerous high-quality papers, and book chapters. He is serving as an Editor and editorial/review board members to many International SCI and Non-SCI journals. He has attracted a few important research grants from industry and academia. His research interests include but not limited to Development of Ionomer/Polyelectrolyte/non-ionic Polymer Nanocomposites/Blends for Smart and Industrial/Engineering Applications.

Dr. Tariq Altalhi joined Department of Chemistry at Taif University, Saudi Arabia as Assistant Professor in 2014. He received his doctorate degree from University of Adelaide, Australia in the year 2014 with Dean's Commendation for Doctoral Thesis Excellence. He was promoted to the position of the head of Chemistry Department at Taif university in 2017 and Vice Dean of Science college in 2019 till now. His group is involved in fundamental multidisciplinary research in nanomaterial synthesis and engineering, characterization, and their application in molecular separation, desalination, membrane systems, drug delivery, and biosensing. In 2015, one of his works was nominated for Green Tech awards from Germany, Europ's largest environmental and business prize, amongst top 10 entries.

His interest lies in developing advanced chemistry-based solutions for solid and liquid municipal (both organic and inorganic) waste management. In this direction, he focuses on transformation of solid organic waste to valuable nanomaterials & economic nanostructure. His research work focuses on conversion of plastic bags to carbon nanotubes, fly ash to efficient adsorbent material, etc. Another stream of interests looks at natural extracts and their application in generation of value- added products such as nanomaterials, incense, etc. Through his work as an independent researcher, he has gathered strong management and mentoring skills to run a group of multidisciplinary researchers of various fields including chemistry, materials science, biology, and pharmaceutical science. His publications show that he has developed a wide network of national and international researchers who are leaders in their respective fields. In addition, he has established key contacts with major industries in Kingdom of Saudi Arabia.